PARTICIPATORY DESIGN:
Principles and Practices

PARTICIPATORY DESIGN:

Principles and Practices

Edited by
DOUGLAS SCHULER
AKI NAMIOKA
Boeing Computer Services
Computer Professionals for Social Responsibility

LEA LAWRENCE ERLBAUM ASSOCIATES, PUBLISHERS
1993   Hillsdale, New Jersey   Hove and London

Lawrence Erlbaum Associates, Inc., Publishers
365 Broadway
Hillsdale, New Jersey 07642

Library of Congress Cataloging-in-Publication Data

Participatory design : Principles and practices / edited by
  Douglas Schuler, Aki Namioka.
      p.   cm.
  Includes bibliographical references and indexes.
  ISBN 0-8058-0951-1 (cloth). — ISBN 0-8058-0952-X (paper)
  1. System design. 2. Human-computer interaction. I. Schuler,
Douglas. II. Namioka, Aki.
QA76.9.S88P38 1993
004.2'1—dc20   92-27297
                                                     CIP

Printed in the United States of America
10 9 8 7 6 5 4 3 2 1

Contents

Foreword

Lucy Suchman
Xerox Palo Alto Research Center

Computer systems development is invariably accompanied by the problem of how to define requirements for the system's functionality. From the developer's point-of-view, the problem has been viewed as one of somehow eliciting from prospective users of a technology what it is that they need the technology to do for them. At the same time this basic problem is often significantly displaced from any specific site of technology-in-use. Imagined users, model users, or surrogate users like the paid subjects of focus groups and operability tests stand in for those who will actually work with the technology. And stereotypic scenarios or extrapolations from prevailing models of generic information processing tasks take the place of an investigation of the specific activities in which a technology will be involved. Even in those cases where development involves extensive inquiry into relevant work activities, it is often persons other than those who actually do the work who speak on their behalf.

Against such a background, this volume takes up the problem of how to establish meaningful and productive interactions among those directly charged with processes of technology design and use. It does so primarily from a designer's point-of-view. That is to say, with the exception of contributions by Ellen Bravo and Frank Emspak, the voices you find here are not those of workers or system users but rather of researchers and developers concerned with bringing their knowledge of technological possibilities to bear on the work of system design. At the same time, what distinguishes this collection from many previous writings on system development is its central and abiding concern for direct and continuous interaction with those who are the ultimate arbiters of system adequacy; namely, those who will use the technology in their everyday lives and work. A key concern throughout is the question of who does what to whom:

whose interests are at stake, who initiates action and for what reason, who defines the problem (or decides that there is one.)

Many of the contributions here are drawn from the first Participatory Design Conference held in Seattle, Washington in the Spring of 1990. That conference was initiated by members of a project on Computers in the Workplace within the national nonprofit educational organization, Computer Professionals for Social Responsibility (CPSR). As a member of CPSR and program chair for the conference I had the pleasure of being part of a unique, grassroots effort to bring this new perspective home to the research and development community within the United States. The conference was, I think, a very special occasion for all involved, distinguished by its deeply cooperative, nonbureaucratic, widely distributed, and extremely effective organizing committee and by the spirit of excitement that characterized the conference itself.

The chapters in this collection follow in the footsteps of a small but growing international community of scholars and practitioners of participatory systems design. Many of the European originators of this approach are represented here as well as some new and distinctively American approaches. At the same time, the contributors vary significantly among themselves in their conceptualization of the processes in which they are engaged. The goal for these writers is not to offer up a general and unified ideology or methodology, but to report in a variety of ways on the problems and possibilities for a more authentically cooperative process of technology design. As a result the collection is characterized less by any single theory or technique that is *Participatory Design,* as by a rich and diverse set of perspectives and experiences that, despite their differences, share a distinctive spirit and a direction. That spirit and direction is characterized by concern with a more humane, creative, and effective relationship between those involved in technology's design and its use, and in that way between technology and the human activities that provide technological systems with their reason for being.

Participatory design makes explicit the critical, and inevitable, presence of values in the system development process. To predominant values of product quality and work productivity are added broadened participation and skill development. The premise is that these values are closely related; that the productiveness of our work is tied to the extent of our involvement, and that product quality is a matter of a technology's support for the continually expanding and developing work practices of skilled practitioners. Some readers may be surprised to find the voice of organized labor represented here as well. Such representation is critical in a discussion dominated to date by managerial and engineering perspectives. Until we become familiar with and take seriously each other's concerns there will be little hope for a mutually satisfactory future in the development of work and technology.

Within this collection, the differences between European, Scandinavian, and United States environments for system development are clear. At the same time, all of these countries participate actively in an international economic community that shares many basic relations of business and working life. To the extent that designers within the United States move ahead with a program of broadened participation in technology development, contradictions will inevitably surface between the rhetoric of user-centered design and employee empowerment, and the realities of hierarchical systems of accountability and control. The point here is not to reconcile these contradictions, but to give readers a sense for the range of the issues. The goal of CPSR, the Participatory Design Conference, and the volume that you now hold in your hands is to widen and deepen our discussion. If as a reader you take away a more sophisticated appreciation for the many meanings of participatory or cooperative systems design, including problems and conflicts as well as prospects and shared values, this collection will have done its job. As a product of our efforts, we hope this volume will be judged by its contribution to a continuing process of discussion, debate, and exploration of alternative approaches to computer systems design.

Preface

Participatory Design (PD) represents a new approach towards computer systems design in which the people destined to *use* the system play a critical role in *designing* it. The approach, pioneered in Scandinavia, is only now beginning to get recognition in the United States. As Paul Czyzewski, Jeff Johnson, and Eric Roberts note in the introduction to the Conference on Participatory Design (PDC '90) proceedings, there are several fundamental ways in which PD differs from *traditional* design.

> It rejects the assumption that the goal of computerization is to *automate* the skills of human workers, instead seeing it as an attempt to give workers better tools for doing their jobs.
>
> It assumes that the workers themselves are in the best position to determine how to improve their work and their work life. In doing so, it turns the traditional designer-user relationship on its head, viewing the users as the experts—the ones with the most knowledge about what they do and what they need—and the designers as technical consultants.
>
> It views the users' perceptions of technology as being at least as important to success as fact, and their feelings about technology as at least as important as what they can do with it.
>
> It views computers and computer-based applications not in isolation, but rather in the context of a workplace; as processes rather than as products.

Participation stands in contrast to the cult of the specialist. In the specialist model, an expert is sought out. The question is presented to the Expert who will eventually produce the Answer. With this approach, those most affected by the conclusion must sit idly by, waiting patiently for enlightenment. PD, of course, demands active participation. PD, however, is not *against* expertise. There is no reason or motivation to belittle the role of expertise. Specialized training and experience, both technical and interpersonal, are important. In the participative model, however, this special expertise becomes yet another resource to be drawn on—not a source of unchallenged power and authority. A partnership between implementers and users must be formed and both must take responsibility for the success of the project.

During the course of editing this volume we have come across several disciplines that prominently incorporate the idea of participation. These include participatory education, participatory architecture, and participatory economics. Although these other participative endeavors are not as familiar to us as participatory design of computer systems, it is clear that these other approaches are motivated by similar concerns as the PD practitioners in this book. The most basic motivation is the idea of democracy. To be more concrete: People who are affected by a decision or event should have an opportunity to influence it. Participation is the key element in democracy. The other compelling idea is that quality can improve with strong and effective participation of the people involved. User involvement and iteration are generally acknowledged to be more critical to success in software design than adherence to conventional design paradigms. Participation is essential to social interaction. It is also essential to good design.

The material here has been chosen in order to illustrate the underlying philosophy and motivation as well as present enough information to begin applying PD in design. We've also sought to transcend the *Scandinavian context* by offering examples and advice that are pertinent to those of us in the United States. To best present the material we have divided the volume into 5 sections. *Context* introduces the motivation for PD in the United States. *Principles and Issues* presents theoretical issues as well as implementation issues. *Approaches to Participatory Design* gives examples of Participatory Design inspired projects in the United States. This section includes descriptions on how to conduct a project using PD. *Guidelines for Technique Development* offers suggestions for how people considering PD can introduce the concepts and ideas to the participants. *Case Studies* talks about the results of PD projects, both in the United States and Europe.

While it is impossible to offer a blueprint for PD, we hope that this volume supplies at least some of the conceptual scaffolding.

ACKNOWLEDGMENTS

We would like to acknowledge the people who have contributed to this book, knowing full well that this list is incomplete. The people listed below contributed to this book or to the Participatory Design Conference (PDC '90) or in some cases, to both.

First, we would like to thank Paul Czyzewski and Jeff Johnson who chaired PDC '90. We would also like to thank Lucy Suchman, who as the program chair was largely responsible for the outstanding program at PDC '90. Lucy invited many of the people from Europe and the United States who have significantly contributed to Participatory Design development. In addition, we appreciate her thoughtful forward to this book.

We are very thankful for all the work that the contributors put into this book. In particular we would like to thank Jonathan Grudin, who not only contributed with two chapters but helped immeasurably in organizing the work, reviewing, and providing an effective European liaison. We are also grateful to the following people for providing invaluable support: Eric Blossom, Darlene Crane, Carolyn Curtis, Marcia Derr, Danielle Fafchamps, Roger Hayes, Heather Holmback, Lesley Kalmin, Dave Kadlecek, Dave Levinger, Max Palevsky, Steve Poltrock, Eric Roberts, Lois Toback, and anonymous reviewers.

Finally, we would like to mention that the conference would have never taken place without CPSR, the CPSR Workplace Project, and the Palo Alto and Seattle CPSR chapters.

Doug Schuler
Aki Namioka

I CONTEXT

1 The Hazards of Leaving Out the Users

Ellen Bravo
Milwaukee, WI, 9to5,
National Association of Working Women

INVITED TALK
PARTICIPATORY DESIGN CONFERENCE
SEATTLE, 1990

At 9to5[1] we're very aware of the hazards of implementing automation without involving the people who will use it. For many years we had a contest called "the pettiest office procedure." The winners one year were lawyers who had installed beautiful new beige carpet in their office. Like the good lawyers they were, they thought about the consequences. They wanted to keep the carpet unmarred, but their secretaries' chairs had rollers on the bottom that would create tracks on the rug.

So the lawyers decided—without consulting or even bothering to inform their secretaries—to have the chairs nailed to the floor. Of course, when the secretaries came to work the next day, they could not perform their job because they had to roll from the typewriter to the computer to the telephone. This is a great example of what happens when you omit the user. Not only does it incapacitate the user, but think about what the carpet must have looked like when the lawyers had to have the chairs unnailed.

I focus here on a very significant and large group of users, namely, clerical workers. A lot of what I have to say can be extrapolated to apply to users at other levels as well.

Clericals are usually omitted from every step of the technology process. They are left out of decisions on how the technology should be introduced, how the equipment should be designed, how the job should be designed. They are also

---

[1]9to5 is an advocacy group for low income working women and families.

omitted from decisions about the application of software and systems. Leaving out the users isn't just undemocratic—it has serious *consequences* for worker health, human rights, job satisfaction, and also for the work process and the bottom line. When called upon for systems design, please consider carefully in what kind of atmosphere and on what equipment these systems will be applied.

Let me begin by talking about health hazards. You're all familiar with the problems of eye strain and vision deterioration that users are experiencing. For example, a 1987 study in Massachusetts looked at 1500 clericals employed at 38 worksites in six different industry groups (Rossignol et al., 1987). Researchers found that 72% of the workers had daily or almost daily eyestrain, about 140% more than people who weren't working on VDT's.

Many studies have shown an increase in temporary near-sightedness. You may have seen an ad that asks what are the three most popular computer accessories. The answer is Tylenol, Ben-Gay, and Visine. What we're finding is that Visine doesn't eliminate temporary near-sightedness, just as Ben-Gay and Tylenol don't take other kinds of problems away. People are experiencing permanent changes in their vision.

Dr. James Sheedy, Associate Professor at the University of California, Berkeley school of Optometry and Chief of the VDT eye clinic there, said they are seeing a higher than normal incidence of focusing problems among people in their 20s and early 30s. Sheedy says many of these people are borderline: They have slight problems that would never appear as symptoms except for the extra load brought on by the way they use VDT's. The problem is not inherent in the technology but comes from the design of the workstations and the design of the job. According to Dr. Sheedy, these problems may actually be causing a breakdown in the focusing mechanism of the eye.

I could tell you horror story upon horror story of what happens to people who work in improperly designed workstations. A California word processor described the various kinds of problems she had: the pain behind her eyes, the headaches, the difficulty reviewing, the blurring when she looked at a distance. She had eye therapy as prescribed by two optometrists—all of which she had to pay for herself, because her employer did not acknowledge her problem as job related. She eventually had to leave and go to a non-VDT-related secretarial job at a cut in pay of $800 a month. Her story is not atypical.

Some problems occur as a direct result of the way the office is set up. I met a women in a VDT training who described how her eyesight had deteriorated a lot over the last 4 years. I asked her to describe what her office looks like. Management decided to set up her workspace as the model office of the organization. They had beautiful wood walls put in with spotlights to highlight the wood. There were hanging strobe lamps throughout the room. And in the middle was the secretary's desk with the VDT. Because her desk had no light on it, the managers decided to train the spotlights directly on the VDT. In addition, they bought her a spectacular wooden desk. She asked for a formica finish, but the managers wanted it to be natural wood to go with the walls. They were afraid the secretary would harm the wood, so they put a pane of glass over top of it. In sum, this poor woman had to deal with reflections all over the place because no one bothered to consider her needs.

The other thing that's contributing to people's eye strain and deterioration of vision is having to read these ridiculous menus in order to pursue any operation on the computer. There must be ways programmers could help to reduce eye strain.

Another big area of health hazards for VDT users is musculo-skeletal strain. We've all seen the pictures of splinted reporters. Once repetitive strain injuries, or RSIs, started occurring among prestigious reporters and even some editors, the big secret came out that RSI is a major problem among people who use computers. A friend who worked in a public defender's office told me how they just plunked down the computers on regular desks. Within 6 months, 4 out of 16 clerical workers in that office had carpal tunnel syndrome.

Some of you should be familiar with this problem from 911 operators in Seattle. When the operators went back to work with carpal tunnel syndrome, some coworkers walked around with their hands curled up and ridiculed the operators for having "puppy dog" syndrome. That's how little recognition there is for the seriousness of what can be not just painful, but a permanently disabling condition resulting from the lack of user involvement in the office set-up and on the job.

The main thing I want to discuss is what has been described as the chief occupational hazard of the computer age, namely, stress. A 1987 study by Denise R. Resko and Phyllis Kernaff Mansfield asked VDT workers and non-VDT workers about a number of symptoms. The researchers controlled for family stress and for job stress that was not connected to VDT's. They found that VDT work in and of itself was creating higher rates of stress symptoms: headaches, gastrointestinal problems, menstrual problems, musculo-skeletal problems, etc. (Resko & Mansfield, 1987).

Why is this happening? 9to5 did a study on stress back in 1984, which was the first time that people really looked at stress among working women. Until then, it had been considered mainly an executive, and therefore a White male, problem. This study, which had 40,000 respondents, showed some very interesting results.

Although women managers were more likely to describe their jobs as being very stressful, clerical workers were more likely to experience the health symptoms of stress. Like the women managers or professionals, the clerical workers had high levels of responsibilities—but unlike the managers, the clericals had little control. It was this absence of control matched with the high level of responsibility that created the stress. The singular condition that most contributed to stress in the 9to5 survey was computer monitoring.

The same kind of results have been found in the Framingham heart studies. When they looked at women clericals, they found they had twice the rate of heart attacks as all other women workers; if you add in variables of low income and children, the clericals had four times as many heart attacks. Why? They were more likely to have an unsympathetic boss, a dead-end job, and no outlet for their anger. In other words, they were more likely to have lack of control.

Considered next are five ways in which technology is adding to stress, then I discuss what that really means in terms of human health and other costs.

First of all, lack of control comes from software designed without knowing or thinking about the needs of the users. A librarian, for example, described a book ordering procedure that originally entailed one form, which now has to be entered on three screens. A dues accounting system for a local union now requires three steps for something that could be done in one step. The person who programmed the system did talk to the staff before developing the program, but did not check back with them afterwards to see how it was working and whether there were any glitches that needed to be changed.

There should be a law: No one who hasn't managed a database should be allowed to program one. I call this law: "No one should be allowed to make menus who hasn't had to eat off them." If you've ever entered data into a database and had to manage it, you know that one of most common things you have to deal with is duplications. You enter a name and then you find that person was already on your list and you want to go back and delete the dup. Say you have two 'Gloria Williams'—if you delete 'Gloria Williams,' you have no "Gloria Williams.' You have to trick the computer: change one of the 'Gloria Williams' to 'Gloria Wilhelm' and then delete 'Gloria Wilhelm' so that 'Gloria Williams' will still be on your list. Why isn't there a simple thing that says: "dup, delete one"? The computer would know there are two; take out one, and you have what you need.

Or how about this for a revolutionary idea. Shosana Zuboff, in her book, *In the Age of the Smart Machine,* (1988), discusses the fear of computer operators that they won't be able to retrieve their data. Computers now work on the principle that what you input is worthless until it reaches a certain stage, at which point you tell the computer: This is now good enough; save it. It should be the other way around: Everything I enter is worthwhile unless I tell you to the contrary. Everything would automatically be saved. You would never have to worry that an error would happen or that the system would go down before you had saved your document. Before you could get out of the document, the screen would say: Name this or type the word "delete." Not: hit "y" or "n." It's much easier to delete what you don't need than to go back and re-key what you lost. The greatest angst, as any one of you who's ever done this know (and I see you nodding your heads), is losing something that you composed from your head with no notes. If the computer treated us as having daily brilliant thoughts, it would save everything we did.

These were examples where the user needs are not understood. The second area related to stress is where the user needs may have been understood, but we'll never know because we can't understand the instructions either in the menus or in the manual. Consider the "stop print" function in some database programs. You make a mistake or you see that it's not printing the way you want and you want to kill it. So you quickly open the manual under "print" and look for "stop" and there is nothing. So you look under "s" for "stop" and there is nothing. So you look under "k" for "kill"; you even look under "a" for "annihilate." And there is nothing in the index that tells you how to make the darned thing stop. Instead you have to pay $135 a year to some user network that you call up when these things happen, and they tell you how to get the thing to stop printing and how to start over again.

I think that there is very little malice involved in people who make these instructions. The problem is people who don't speak conversational English, even though they are not foreign-born. They simply speak "tech-ese"; we need an interpreter to understand them. It would obviously be a lot easier if the people who used the system helped design the instructions.

A third area that contributes to stress is software designed without regard to the impact on the user. An insurance clerk told 9to5 that every time the system was about to go down, a message flashed on the screen saying: "fatal error." Even when you know that you, in fact, have not murdered anyone with your keystroke, the stress effect on your heart and breathing are the same every time the message comes up.

TWA has a new call distribution system. It has eliminated the 6 seconds that you used to have between calls to finish scribbling your paper work, or take a sip of water, or maybe crack your neck. There are now no seconds between calls. Clearly, whoever designed the system had no idea what it would feel like the instant you hang up to have to pick up the phone again.

The fourth area is software that is designed to manipulate or control the user. On an episode of "L.A. Law" Roxanne led the secretaries in a walkout. One of the issues they were up in arms about was VDT's. They were concerned about the hazards, but they were also really angry because their machine was flashing messages saying, "My world is calm," "My world is productive," The secretaries were incensed. Their world was not calm. They were trying very hard to be productive. They found the messages patronizing and offensive. Nowadays software manufacturers boast about programs with this capability. Greentree Publishing, for example, advertises a software package called "Subliminal Suggestion and Self-Hypnosis" with everything from feel-good messages to more pointed commands like "work faster."

But the fifth area related to stress is the most insidious, the fastest-growing, and the most frightening: The computer is actually used to spy upon the user. We tend to think of spying as something you do to the enemy. And we treat very seriously people who spy on us as a country or who sell our secrets to other countries. Yet we are promoting the use of technology as spy and the worker as the enemy. Here the worker is no longer in any way a subject but becomes the object of the system.

Imagine being a reporter writing your first draft. You're just kind of spitting out your thoughts. Suddenly you get a message on the screen from your editor: "Bad lead. Start over." PC Week Magazine advertises software with such capability by telling the reader: "Look in on Sue's computer screen. You monitor her for a while. . . . In fact, Sue doesn't even know you're there."

In a New York company that had 100 or more terminals in one room, someone told us that occasionally this message would appear on her screen: "You're not working as fast as the person sitting next to you." The computer is able to count every keystroke that you enter, every minute that you are away from your desk to use the bathroom. In fact, it is now not uncommon to have bathroom break statistics posted, with the people who need the fewest potty breaks minutes at the top of the list. This is going to give a whole new meaning to "wellness program," a whole new set of exercises to learn to keep from going to the bathroom outside of break.

Let me share some very serious examples. A lot of this happens in the airline industry. On our job problem hotline we heard from an airline reservationist who was suffering from severe mental stress, insomnia, and stress-related jaw spasms. This is how she described her job: "The computer tracks your time on the calls and it's connected to a light on the tower. If the light goes out for more than 14 seconds, they come on and listen and make nasty comments at you through the earphone."

Another sales agent described how workers punch into three different things: the VDT; a keypad called the Collins; and headsets. Management tracks every second of their working day. Daily and weekly printouts map the average call length, the length of time between calls, the unplugged time, and sales performance. This woman said, "I get more grades in a month than my kids get in their whole life from school." Supervisors can listen into your calls not only with customers but with your coworkers. And you get demerits if you have too much unplugged time, if you didn't sell enough rental cars, if you didn't get enough people to get their tickets in the mail, etc.

This woman, although she'd been there seven years and had a good work performance every year until they implemented this system, was threatened with losing her job because she had too much unplugged time—read: time in the bathroom, time between calls. And she was very nervous, trying to get her statistics up. Her supervisor noticed she didn't look well and told her, "Take 10 minutes and pull yourself together." And the woman said, "I didn't know what was happening to me, but I had a feeling it was going to take more than 10 minutes to take care of it." In fact she had a nervous breakdown and was in therapy for 8 months.

Practices like these have very serious consequences for our whole concept of worker rights and the issue of worker health. A group in Massachusetts did a survey on the consequences of computer monitoring. Sixty-two percent of the people they interviewed did not know at the time they were hired that they were going to be monitored on their job. Barbara Garson (1988) in *The Electronic*

Sweatshop gives an example of a woman who plugged in her headset, turned to a coworker and said, "The doctor says it's cancer," then took her first call. An hour later her supervisor called her in and said, "Is there any way we can help you with your cancer?" The woman stormed out of the office, marched up to her coworker and said, "How dare you tell them." The coworker replied that she had said nothing. That was the first time they realized that their private conversations could be overheard by the same people that were listening in to their calls with customers.

Almost one-quarter of the people who answered the survey in Massachusetts said that their individual stats were publicly posted either by the week or by the month. This misuse of computer systems, treating the user as the object rather than the subject, has serious consequences for worker health and also for the bottom line. A friend of mine did a study on the introduction of technology into three different workplaces and what its impact was on users. She interviewed 161 clerical workers and asked them various questions about how automation had been introduced. The two conditions that most correlated to a high rate of health symptoms were too little training and too much down time—where the user, in other words, had too little control to use the machine effectively.

There are lots of studies that show that productivity goes up 10–25% when you create healthful and ergonomically designed workstations, and the opposite is true when you don't. Harley Shaiken has written about studies demonstrating that computer monitoring is counterproductive. People say they use this technology to improve productivity, but it doesn't work. It makes you wonder if that was ever the intent at all, or if in fact the intent, as Shaiken maintains, isn't rather to control and intimidate workers.

I've been sitting on a commission in New York on the skills for the workforce 2000. They have identified two different trends in management. One trend, representing about 5 to 10% of businesses, is toward a high skilled, high value-added workforce, which requires a new form of work organization—what you refer to as "participatory design," a team approach. But the majority of work-places instead have a low-skill, low-cost, high turnover view of management. If you have that view of how to do business, obviously you need control in order to get your workers to do the job because the attachment, loyalty, morale don't exist without investing in the workers.

We all know Big Brother didn't work in Eastern Europe. Why is it taking some people so long to figure out that it's not going to work in the United States? It's going to take a lot of work on our part to see that it stops. In answer to the question, "How do we get management to realize that these things need to change?", the most significant way is by creating the power among workers and like-minded professionals to make them change.

You can also use arguments of the bottom line. Stress claims cost an average of $11,000 as compared to $5,000 overall for worker compensation claims. The National Institute of Mental Health said that 30 million Americans suffer mental strain on the job that is workplace-related. A third of those people are under the age of 30. More than half of them are female. We are talking about symptoms that may affect a minority of workers, but they affect them in very significant ways, both for the workers and for the workplace.

Clericals have been thought of as *just users, just women, just secretaries.* The idea that they would have something to say about how the work should get done is really a radical departure from business as usual in the United States. We also have management that focuses on short-term rather than long-term cost effectiveness. It's not really more expensive to do what we want to do, but it's not the way that American management on the whole thinks. And it's clearly a question of power.

Pay attention to this author who characterizes much of American management style as the "military method of leadership." The author asks why so many American managers ignore the basic rights that are enshrined in the Declaration of Independence and then gives this answer: "Simple. It requires the CEO's and the top VP's to give up power. This is threatening to them." The author of that opinion piece, by the way, was not a radical 9to5 member, but the president of Catholic Knights Insurance Society in Milwaukee, a man named Daniel Steininger (1990). He also quotes some remarks from the man who started quality control circles in Japan, to the effect that companies exist in order to guarantee a good income to their employees and to see that they live a happy life; and if they don't carry out this function, they don't deserve to exist.

Clearly, the only way to reverse this situation is to involve users in decision making and that means some radical re-thinking about how jobs should be designed. Variety and rotation of tasks, multiskill approaches have to be incorporated into the actual job. There are people trying right now through various approaches to make this happen. At Ohio Bell, for example, a collective bargaining agreement limited secret monitoring. It said management could take aggregate data, but could not use data on individuals. People have been trying litigation with mixed results. There are four key cases, two of them in favor of the worker, two of them in favor of management. At 9to5 we are looking for a test case linking the issue of monitoring with the right to privacy. And legislation has been introduced, the Privacy for Consumers and Workers Act, which provides mainly for the right to know and privacy protections.

We need to see a much more interactive model, where users are involved at each step of the way, where the locus of control shifts to include or mainly be in the hands of the users, where everything starts from the needs of the workers. If workers don't yet have the ability to conceptualize what they need, they certainly know a starting place. We need to figure out how to create the language and the interaction so that the conceptualization can take place with the users playing an integral part. We need to have designers experience work so that they know how the application is going to be used and what it will mean.

I end by cautioning you about the concept of participatory design. The idea of worker participation is being espoused by some people in top management, but I suspect in very different ways. There is a big difference between making suggestions and making decisions; and there is a difference between having the right to participate and having power. As you pursue the issue of participatory design, I encourage you to make sure that users include clericals, and to make sure that participation involves decision making.

Editor's Note

More information on the topic of computer monitoring and health and safety issues can be found in two 9to5 publications: *Stories of Mistrust and Manipulation: The Electronic Monitoring of the American Workforce,* and *VDT Syndrome: The Physical and Mental Trauma of Computer Work.* To order, call 9to5 at 216–566–9308.

REFERENCES

Garson, B. (1988). *The electronic sweatshop: How computers are transforming the office of the future into the factory of the past* (p. 69). New York: Simon and Schuster.

Resko, D. R., & Mansfield, P.K. (1987). Video display terminals: How they affect the health of clerical workers. *American Association of Occupational Health Nurses Journal, 35*(7), 310–314.

Rossingnol, A. M. et al. (1987). Video display terminal use and reported health symptoms among Massachusetts clerical workers. *Journal of Occupation Medicine, 29*(2), 112–118.

Steininger, D. (1990). Workers' happiness should come first. *Milwaukee Journal,* March 23, p. C11.

Zuboff, S. (1988). *In the age of the smart machine.* New York: Basic Books.

2 Workers, Unions, and New Technology

Frank Emspak
School for Workers University of Wisconsin—Extension

The best way to change the structure of American industry in a manner that preserves productive employment, produces a better quality product, and restructures the technology among more democratic lines is for users and designers to find the means to talk with one another as equals and then actually effectuate a decision. To be able to talk with each other as equals, to engage in a participatory design process, we need to construct a support mechanism to enable the process.

In this chapter, we focus our discussion on how people in the factory see new technologies being implemented. In effect, we focus on one class of user. We emphasize metal working and factory work primarily because unionized workers have been the most able to express concerns about the way their work is structured. However, the issue of top down implementation and design of new technologies is the same for office workers, hospital workers, as well as draftsmen, planners, and large numbers of lab technicians.

Workers at all levels share the problems of the apportionment of skill, the design of our tools and equipment (especially software). It is, therefore, apparent that we need to forge links with all users and thus develop more global criteria for the construction of software systems, factories, and the productive apparatus including all office work and data processing.

Although unionized workers in the private sector have traditionally been the ones to raise questions about technology implementation this may be changing. Organized public sector workers may soon be in the best position of all to raise questions of work organization, technology, and enhanced quality of service resulting from meaningful worker input. The demand to cut costs in the public sector may provide unionized workers with a means to push a worker-oriented skills-based agenda. When employees have a chance to reorganize work and

13

make efficient use of new technologies the organizational structure of the system is altered reducing many expensive and not necessarily productive management positions while maximizing expanded service delivery roles. As more and more office workers get organized, particularly in the public sector, and express themselves collectively, we anticipate increased discussion of job restructuring in the office setting.

Throughout the chapter, I will use the terms *worker* and *labor*. Who is the *worker* and what do we mean by *labor* in this essay? The *worker* is defined as the nonprofessional work force. *Labor* tends to indicate blue collar organized labor. Generally, the old terms blue and white collar, with the exception of professionals, delineate this group of workers. The mental paradigm is the traditional factory. In most such institutions the line between those who implement and those who program and design is very clear.

Ultimately, any strategy that has as its objective bringing together conception and execution will have to involve computer programmers and designers. At some point, workers, programmers, and designers will have to begin to meet together and figure out how to work collectively in order to bring about change on the shop floor. In an addendum to this chapter, I suggest a model for a technology design project which may provide a framework for cooperation. However, the chapter itself is more limited. My goal is simply to discuss how *technology* is viewed looking up from the bottom, and suggest steps that organized labor can take, in the immediate future, to enhance its ability to represent the work force.

A FAMILIAR STORY

The story is familiar. Workers notice some new managers and engineers in the work place. People are measuring something. All of sudden new equipment or computer terminals come in. Some individuals are assigned to the new system—but no one explains how the new system *really* works. Different jobs are created—but it seems that management keeps the interesting work—even though workers used to do it. Older workers aren't trained on the equipment. All of a sudden they aren't flexible enough and don't know enough. New people usually brought in out of seniority are put on the job.

The union leadership has been told some new equipment is coming, but they have not been told much about its capabilities. The leadership is uneasy—and so are the workers. But after a few weeks of talking with management about the effects the new equipment may have on the work force, only some details like shift assignment and some bumping rights are straightened out. There is no discussion about the new jobs and management is still doing the work. No one feels quite right. . . There is a sense of loss.

Often the introduction of new machines or office procedures is accompanied by a veiled threat. Either do it the new way or no new investment. In these times of plant relocations and shut downs only a fool would argue about how investments are made—at least so the story goes.

Besides—who knows if there is a better way to put in new technology. After all, don't the engineers and management know best?

The story does not end yet, although the *official* role of the union usually stops after the conclusion of the first flurry of discussion about new means and methods. But time goes on and the work force including management tries to get used to the new equipment. To a degree it becomes part of the landscape.

But the equipment is not working too well. Even after a few months the new system is not living up to expectations. Start up costs are high. Problems happen all the time. Quality problems show up and the company is still in trouble—threatening their own existence as well as our jobs.

Is There Another Way?

There are other ways of implementing technological change that allow workers to enjoy the promise of the new technologies. To estimate if those ways can be of use to American workers we must first understand what technology is and who actually implements it.

What is Technology?

A good definition of technology is *the means and methods through which we as a society produce the substance of our existence.* What does that mean? There are old tools, like a shovel and new tools like a robot. There are old sources of energy like steam and new sources of energy like electricity. Materials are also part of technology. Old materials like cast iron and new materials like fiber optics. Then there are techniques of production—old, like weaving, and new as x-ray lithography for producing micro chips. Finally, there is the organization of work—craft work is an old form of organization, the assembly line is newer, and various computer aided flexible manufacturing systems are the newest.[1]

WHO IMPLEMENTS THE TECHNOLOGY?

In spite of massive changes in the technical base of our factories and offices it is still the worker who implements the technology. As yet there is no substitute. On the surface there appears to be a contradiction between the tremendous investment in sophisticated equipment and its use by relatively unskilled workers.

---

[1]Since the beginning of the production of goods, there has been an *old* technology and a *new* technology. The pace of change, however, has altered.

Sometimes this contradiction expresses itself as the new jobs are more boring and less creative than the old. However, the fact remains that people are needed. And because people implement even the most costly and modern technical system there is a way out of the seemingly downward spiral of job displacement, poor quality, and excessive costs associated with the introduction of new technology.

The way out derives from the concept that most working people have the same aspirations as everyone else. Increasingly, working people and their organizations are coming to realize that the technologies that are being implemented in the work place are not immutable and can be changed. The introduction of new technology is an opportunity for labor to regain the initiative in regards to management over job content, skill, wages and benefits.[2]

TECHNOLOGY: THE PACE OF CHANGE

For about 50 years from 1920 to 1970, the technology in industry changed incrementally and relatively slowly compared to the last 10 years. The basic job structure was designed by F. W. Taylor, who proposed reducing each individual job to its smallest least skilled portion as a means of achieving management control and hence efficiency.

Over time the basic trend in work organization was to separate conception from execution. This reached its most capital intensive form in the automobile industry although much of the so called *women's work* in light manufacturing, and almost all office work is organized in the same manner. In the early 1950s, the first numerically controlled machines were designed. The Numerically Controlled (NC) machine tool provided a means to further separate execution from conception. At that time some firms specifically identified deskilling workers to reduce control over production as one driver behind the development of new technologies—specifically the numerically controlled machine tools.

Starting in the 70s the technological pace of change picked up. Not only did technique change but the actual job structures changed as well. The organization of work changed. In the machine trades the programmer increasingly decided on to speeds, feeds, and tool sequence. These functions had been in the heads of the machinists. In large manufacturing plants fairly rigid production control systems like Materials Resource Planning (MRP) displaced the clerks who had tradi-

---

[2]But in order to take advantage of this opportunity working people need some ingredients which have previously been missing in a view of technology. First, working men and women, and our unions, need to understand the new technologies with the same precision that we understood the older technologies. Second, from that understanding workers need to project a vision of what is possible. Above all working people need to regain and build on their self-confidence. The basic building block of our self-confidence is the pride which most workers share in producing a good piece of work.

tionally specialized in production control functions. To a greater and greater degree people became monitors rather than doers.

Over the years the response of organized labor to issues of technology has centered on the effects of technology, not its design. When industrial unions organized themselves in the 1930s the structure of work was not challenged by the CIO and, by and large, was not an issue. By the mid 1970s the pace of technological change was fast enough so that many union staff people, full time officers, and workers with seniority active in the union but not affected by the *new* jobs lost their understanding of the actual way in which work was performed. The net result was that unlike European unions, which are more involved in issues of job content, American unions until recently did not see this as a major issue.

THE BARGAINING PROCESS: DOES IT FIT THE RAPIDLY CHANGING CONDITIONS?

The essential outlines of modern collective bargaining became law 50 years ago. In the bare essentials industrial unions agreed to negotiate about the effects of the technologies. The legal structure that regulates unions specifically restricts unions from having a right to bargain on "means and methods." (Management may agree to discuss the issue, but it is not a mandatory subject for the purpose of collective bargaining.)

Management has the right to choose tools, energy forms, techniques, and organization of work. Unions negotiate primarily concerning compensation— including benefits. An important exception has been the increased activity regarding health and safety effects of the machines and office equipment. One group of unions—the craft unions maintained their practice of intervening concerning means and methods. They have been able to continue this tradition through the control of training, especially apprenticeship training.

This bargaining structure no longer meets the needs of many workers. Bargaining that is limited to wages, hours, and a narrow definition of working conditions in a rapidly changing technological world fails to ensure a better standard of living for most workers. However, what choices do workers have? Having to change the bargaining system in order to negotiate concerning technology hardly seems like an opportunity. Labor law is fossilized and administrative law mechanisms are controlled by business interests seemingly dedicated to marginalizing unions. But there are forces in the manufacturing world that recognize change must occur.

FORCES FOR CHANGE

If only union members were negatively affected, then the chances for reform of the labor relations structure would be very poor. Research and experience indicate, however, that it is also true that the modern equipment does not work as effectively, quickly, and at as high a quality as possible without the in depth and voluntary participation of the work force. Although management resists many changes that limit its authority, American firms also have to face competitive pressures abroad. Successful competitors have been those industries that have been more willing to change the scope of collective bargaining and bring labor to one degree or another into the process of determining means and methods. American management may not like it but if faced with a determined well organized union projecting a program of how to get things done better, bit by bit the door to more innovative and inclusive agreements would be opened.

What do we Need to Know?

The best negotiating committees are formed by members who understand the contract and who also have a detailed knowledge of the equipment they work with. They can use this knowledge to bargain effectively. When that knowledge is diluted or removed from the purview of the union its ability to bargain intelligently on any issue is reduced.

Understanding *technology* means a basic understanding of the work process and its potential. Unions, and workers in general, need to know what software is and how it can be programmed. Working people need to know that software can be redesigned so that workers on the shop floor can program it. Users need to know if *one way* information systems often found in offices are truly one-way or if they can be reconfigured for increased operator interface.

Part of our understanding must revolve around the issue of worker intervention. One must know how flexible the systems are—and as part of that one must know with what precision systems are capable of monitoring work.

Understanding the production processes, including work flow, scheduling, and the operation of the firm's information system allows the local union to bargain effectively because it can control production. By understanding the real position of the worker in the firm's productive apparatus the union membership can plan for the future. For example, if the specific jobs in question are transitory due to technological change, for example, key punch operators, then emphasis has to be put on training and alternative job paths for those workers. Likewise, as working people better understand how new technologies impede or expedite the transfer of production to non union areas labor's legislative and bargaining objectives can be altered to match those dangers more precisely.

TECHNOLOGY IS NOT A SEPARATE ISSUE

Technology bargaining is not a separate issue from other parts of the contract. It should not be treated as one item on a laundry list of items. The companies understand that the design and implementation of means of production are central to the work place and so should unions. Firms usually design their bargaining strategy based on how they want their equipment used and don't treat *technology* as a separate issue. Neither should unions.

Technology bargaining could be called bargaining over means and methods. Trade unionists can use such bargaining as a way to project a vision of the workplace as a means of motivating people to do something about it. Very few people get all excited about a grievance or some technical change in the contract. But there is much more interest when unions start talking about more general ideas including adding some skill and control to the workplace.

The Workers Role in Job Design

What should workers do if their firm is willing to negotiate the design of a new system? The first task is to create a list of design criteria. These criteria should reflect the reality that the firm must be profitable. Thought must be given to criteria that need to be added for the system to maintain employment. Criteria need to be defined which can enhance skills. The second task is to reject criteria based only on minimizing jobs and employment. If the firm is to increase market share, produce better, etc., then overall costs have to decrease. In fact, based on our earlier criteria, costs of the hourly wage for our members may go up in absolute terms—although maybe not as a percentage of the cost of production. However, overall costs of production may go down as quality improves and as the firm is able to respond more quickly to problems, as a result of increased skill and flexibility within the work force.

At this point the problem is not a technical one. It is a political, technical, and economic one. Workers have to convince management to bargain about how they want to run the enterprise. Can this be accomplished?

The situation, legally, is that management has a technology rights clause. They also have a right to do everything else clause. These clauses are called *a management rights* clause. It is unlikely that a union can insist that this be dropped as a precondition for discussions. But if workers are successful in getting management to the table then in the next round of formal contract negotiations they should consider trying to limit this clause.

In order for workers to get management to talk, there are two tracks that can be pursued at the same time. The first track challenges management as to their overall strategy. In other words ascertain if they just want to reduce labor costs or do they want to increase productivity and quality?

In order to challenge management workers need to have time. Part of negotiating about technology is to have time to analyze the new systems. It is important to insist on formal, early notification of any impending changes, along with the company's estimate of the impact on the work force. Meanwhile, it is also important for workers and the union, using their own sources, to have an independent estimate of the expected effects of the new systems on employment and skill.

Workers can pursue another track too. They can learn how the office or plant works. A union local needs to have some ideas (even if they are not completely right) as to how to run the enterprise better. It could be, for example, more training; it could be worker programmable equipment; it could be another dispatch system or production control system. In other words working people collectively have to know as much about the plant as individual workers know about their job. Jobs are too important to leave it up to management.

What happens if that strategy idea does not work? At least by pursuing track one workers probably know what the company is going to do. If a union directly offers management a willingness to discuss change and better ways to produce something, management in generally divested of its ability to go to the union members and accuse the union of standing in the way of progress.

If the union has done its work, it has used management's answers combined with its own research to provide the membership with good reasons for why the union didn't buy the company's approach. Hopefully, due to the cohesiveness of the union and its understanding of the productive process, it may convince the firm to rethink its position and bargain.

Assuming that a firm is willing to bargain, how does one use the ideas of participatory design to save jobs or enhance skill? At this point the task may be more technically difficult, but a union has made an important organizational gain. We have to put the union's criteria for machine design into practice.

Often a firm has already defined its design objectives, and may be ready to order specific equipment. It is in the union's interest to independently evaluate the system or machines based on the union's criteria as well as the firm's.

The next step is to investigate alternatives. There are several approaches to this step. Three of them are:

1. Use the resources of the international union to get an idea of what type of machines or systems are on the market, or how other locals have dealt with the same problem;
2. At the company's expense visit machine tool or system designers, local technical universities, etc. and compare equipment;
3. Work with professionals with expertise in human factors to make sure you understand the alternatives.

What work are we really doing by searching for alternatives is acquiring equality of knowledge. If one is ignorant of what alternatives are possible, then it will be impossible to win them. The central point in making ideas into reality is knowledge.

The Center for Applied Technology

There have been some limited examples of worker participation in designing production systems. As a means to encourage modernization of the manufacturing base, the Commonwealth of Massachusetts organized a project called the

Center for Applied Technology (CAT). One means that the center employed to bring about change in the work place was to encourage the formation of joint worker management teams to solve production and quality problems.

The experience of CAT showed that it is not necessary to have all the answers—but only necessary to pose alternative ideas of manufacture. CAT reached people in management who thought about production and was able to help them organize joint worker-management projects that discussed design and implementation issues.

The CAT succeeded in going through all the steps noted above at the Pneumatic Scale company. Local 444 of the Retail, Wholesale, and Department Store Workers Union originally interested the firm in the CAT program and consistently encouraged the firm to stay involved in the project. The firm employs about 350 people engaged in the machining of parts and their assembly into packaging machinery, mostly container cappers and fillers. Customers include Proctor and Gamble and Lever Brothers.

CAT assisted in organizing a team including the president of the local, the vice president, the manager of industrial relations, and the manager of production to work out details. Every so often all the top management of the firm would come for a report and agree, disagree, or in consultation with CAT and the union, modify the proposals. Then management would implement.

The most significant part of the program was the union's role as initiator. It came to the CAT and brought management with it. The union insisted on revamping the training. The union pushed for changes in production. Noteworthy aspects of this approach are the equality of knowledge, the scope of the decisions that the team was allowed to make, and the fact that the union was represented in its own right.

THE VISION

A new approach that has proven its value is skills based automation. Another way of saying the same thing is *user driven design*. Both concepts take as their starting point that the user is the worker—not just the management. Generally speaking, when the machine tool and computer people talk about users they combine the buyers (management) with the actual user (workers). Their designs conform to the financial and organizational needs of the firm first, not necessarily to the needs of the production people. Taking it one step further, user driven design places the needs and abilities of the worker at center stage along with the other needs of the firm.

Skills based automation as a goal and negotiating tool allows working people to take the offensive in regards to technology implementation. It also allows organized labor to move off the defensive in the discussion of *productivity, competitiveness,* and training. It allows labor to take the high ground when it comes to flexibility. The most skilled workers are the most flexible. A skills based conception of work enables working people to devise a manufacturing system that produces a *quality* product in a cost effective, efficient manner.

Can Organized Labor Project a Vision of Change in the Interests of all Workers?

The ability of the labor movement to defend all workers by projecting a positive vision is a necessity if the labor movement is to break out of the isolation imposed on it by the corporations. A skills based view of work allows organized labor to set the agenda and defend the interests of all working men and women, not just a few affected by a specific process. Both professionals and nonprofessional workers are interested in jobs that include a continuous self-improvement factor and during the course of the day combine physical and nonphysical elements. Most people want jobs that offer some creativity and responsibility. Almost all working people want to have some say in their pace of work and not be machine paced or computer driven.

Working people need to be able to recognize that technology is not some sort of force outside of their lives, but a socially determined system. Like any other socially determined system a technical system reflects certain values and assumptions. These values and assumptions may favor management at the expense of workers. Most unions and working people understand this when it comes to questions of health and safety. Raising the issues of design and implementation is not much different than raising issues regarding safety and health.

The labor movement has at its disposal a great reservoir of information regarding new technologies. This knowledge resides in the membership of the unions and in the many unorganized working men and women who would be only too glad to share their knowledge and expertise with their fellow workers. It is probably accurate to say that the knowledge potential of the trade union movement far surpasses the information available to management. So the basic ingredient for participatory design knowledge is present. The only ingredient lacking for the moment is the vision to use it.

Addendum

It is not enough to suggest that working people can and must participate in designing the technologies used at work. We must outline the forms within which this process can take place. The School for Workers, a program of the Extension, University of Wisconsin, has proposed a technology assessment and design project in the fine cabinet industry. The model has the potential of being applied in many industries.

The project can encourage meaningful participation by workers, engineers, and designers, as well as progressive firms in the industry. Although we in the

United States do not have the institutional support that could assist participatory design efforts—as do our European counterparts, we think their model has a great deal of relevance. Thus any resemblance between the project we suggest and Scandinavian participatory design ideas is deliberate, and we thank our colleagues abroad for their pioneering work.

PROPOSAL

Skill Based Automation
For the Fine Cabinet Making Industry

Objective

The application of modern computer aided wood working equipment and skill to the fine cabinet making industry will allow the industry to manufacture high quality, one-of-a-kind products at competitive costs. The results will include an expansion of the industry while the quality and variety associated with the industry is enhanced.

The skill composition of the work force will also change. Of necessity there will be a process wherein the traditional wood working skills will be combined with newer skills associated with flexible computer aided manufacturing. The result will be that an enterprise will have a work force that combines the new and the old at a higher level.

The project will demonstrate in a practical fashion that a combination of traditional cabinet making skills combined with flexible automation can:

1. produce high quality fine cabinetry at a reasonable cost;
2. preserve and enhance the skills associated with the trade;
3. indicate the value of a union environment as opposed to a nonunion environment.

Phase One

Project Organization.   The Carpenters Union and the University of Wisconsin School for Workers will organize a project leadership body. The body will include individuals from the Carpenters Union national staff, working cabinet makers, local officials, and faculty from the school. Members from the industry will also be asked to participate—either directly or as part of cooperative agreements that exist between the Carpenters and the owners. The chairman will be from the Carpenters Union and the cochairman will be from the School for Workers. It is understood that the committee will be answerable to the Carpenters Union in all matters.

Duties of the Committees.   The committee will direct the skill based automation project as defined below. The committee will also evaluate the results of the project, set criteria that direct the design of machinery and software, and in general oversee the project in all its detail and complexity.

A brief outline of Phase One follows.

1. *Organize a project leadership committee.* The School for Workers will assist in developing a joint committee consisting of Union and management representatives.

2. *Convene the committee.* Assess whether current problems in communications exist between the worker (union) and management representatives, using the facilitation services of the School for Workers.

The School will provide the necessary training in problem solving to enable the participants to work together on a different and positive basis.

3. *Develop goals for the project.* The School will assist the body in reaching a consensus as to the specific goals of the project. The committee will jointly draft the objectives and implementation strategy.

4. *Educate the committee.* The School will provide the committee with a full explanation of the concepts of skill based, flexible automation. In addition, we will review the current technologies available to the cabinet making industry.

5. *Define criteria.* Once our education is complete the committee will define the criteria by which we will judge the hardware and software available for use in our industry.

6. *Identify skills.* The committee will identify key management and working skills, essential for making best use of the technologies.

7. *Identify resources.* The committee will identify engineers, software designers, and machine tool manufacturers who are best able to advance the objectives outlined in the project.

Phase Two

Two distinct problems must be addressed if we are to use new, programmable machinery to produce fine cabinetry. First, appropriate hardware must be identified, or if unavailable on the market such hardware must be designed. Second, the best software that governs the tool movements must be identified. If software that matches the reasoning process of the cabinet maker is not presently on the market, we must design and produce the software.

1. Hardware Identification.   A survey of existing wood working equipment will be conducted by the committee or its designee. Equipment presently in use in the industry will be rated according to criteria developed by a Carpenters Union study committee.

2. Software Identification.   A survey of exiting software will be conducted by the committee or its designee. Software presently in use in the industry, in this country and abroad will be evaluated based on the criteria developed by the committee. While members of the committee may be designated to participate, engineers or other persons with expertise including working cabinet makes, will also be solicited to identify needed software.

Phase Two will be complete when the software and hardware is selected and rated according to our criteria.

Phase Three: New Design

In all probability the evaluation phase will discover that existing designs, especially software designs, do not match the needs of the work force, or of the industry. If such is the case the committee will evaluate the costs and benefits of modifying existing software. If our analysis is positive the committee will commission the design of new software to meet our needs.

Phase Four: Implementation

The committee will arrange for a demonstration project of the hardware and software defined above.

The committee will direct an evaluation of impediments to the organization of cabinet making using the production system that we as a committee have identified. Our evaluation will include an assessment of the application of human and management resources to the changed work place. Based on the findings of that evaluation, appropriate educational and organizational steps will be implemented in shops that are willing to be used, in whole or in part, as implementation sites.

Phase Five: Dissemination

The results of each phase will be disseminated to participants as soon as practical.

Editor's Note

Frank Emspak is an assistant professor in the School for Workers, University of Wisconsin Continuing Education Extension, Madison Wisconsin. Prior to joining the School, he was project Director of the Center for Applied Technology, a state funded program headquartered in Boston MA. Before coming to CAT, Dr. Emspak was a machinist employed by the General Electric Company, Wilmington MA. While employed at GE, Dr. Emspak served on the Executive Board of his local union IUE Local 201 (AFL-CIO) for 8 years.

FURTHER READING

Ahlin, J., & Svensson, L. (1980). New technology in mechanical engineering industry: How can workers gain control? *Economic and Industrial Democracy, 1,* 487–521.

Ashford, N., & Ayers, C. (1987). Changes and opportunities in the environment for technology bargaining. *Notre Dame Law Review, 62*(5), 810–858.

Bernard, E. (1987). *A union course on new technologies, 14,* 3–4, University of Waterloo, Ontario, Canada.

Brodner, P. (1985, September). Qualification based production—The superior choice to the "unmanned factory." *Preprint from the 2nd IFAC Conference on Analysis, Design, and Evaluation of Man Machine Systems.* Varese, Italy.

Cooley, M. (1989). European competitiveness in the 21st century: Integration of work, culture and technology. *FAST Action Program of Human Work in Advanced Technical Environments* (Final Draft Report 27), June 1989.

Martin, T., & Ulrich, E., & Warnecke, H. J. (1987, July). Preprints. *10th World Congress on Automatic Control.* Munich, Germany.

3 A Design of One's Own: Towards Participatory Design in the United States

Joan Greenbaum
La Guardia College, City University of New York

The road toward participatory design in the United States has been under construction for some time now. Certainly by the middle part of the 1980s as most large companies grappled with the proliferation of PCs on employees desks, it was becoming clear that users of computer services needed more of a voice in the type of computer services they were receiving. This chapter lays out some of the groundwork that has been leading to more user-centered system design, and explains how this can be used as a bridge toward building more participation into the system development process. Several issues emerge here. The first issue is that user-centered design, while helping to pave the way towards participatory design, is not participatory in and of itself. The second issue addresses the differences between what has come to be known as the Scandinavian approach to design and the way participatory design in the US is emerging and is likely to develop. The third issue is that while participatory design both requires and fosters workplace democracy, participation in the design process may not necessarily lead to workplace democracy. In fact, a central point of this chapter is to illustrate the ways that, I believe, American design needs to take on its own *home grown* characteristics and to paraphrase Virginia Woolf, become *a design of one's own*. As the chapters in this book point out, participatory design is many things to many people. Yet there is a remarkable core to the ideas which have been built on common ground. Among the elements in the common core, are the ideas that: Computer applications need to be better suited to the actual skills and working practices of the people using the systems; that work is a social activity involving the interaction of many groups of people; and that barriers between technical specialists and people using computer applications need to be broken down in order to build effective communication during the design process.

Participatory design implies that workers as users of computer products should take part in the decisions that affect the system and the way it is designed and used. Because technology is not developed in isolation, participation in decisions about technology also involves decisions about work content and job design. Seen in this broader context participatory computer system design needs to be part of an *integrated design* that looks at work organization, job content, and the way technology is used to support these activities. This integrated design process also implies that system developers as technical specialists need to refocus their energies so that they can learn to rely on the expertise of the workers. System developers in an integrated and participatory environment would then take part in playing active roles in fostering and enabling people to use their knowledge to make decisions. But the step between fostering participation and enabling decision making is grounded in a question of power.

POWER SHIFTS: CONTROLLING THE SYSTEM DEVELOPMENT PROJECT

Participatory design represents a potential break with traditional approaches to system development. The break happens when we recognize that full and active participation of users as workers implies a shift in power relations within companies. To address this issue more closely the chapter focuses on Information Systems (IS) within large organizations, particularly Management Information Systems (MIS). In the development of MIS applications the contradictions between the roles of system developer as management consultant and system developer as spokesperson or catalyst for user groups become quite clear. System analysts, whether they are employed by an organization ordering a system or work as outside consultants to a project, are usually hired by management to respond to a set of management defined objectives. Within the American context, the idea of a system analyst acting as an advocate for user rights is not yet an issue on the table of possibilities.[1]

This means that when system developers take an active role helping users become involved in defining and using computer systems they are moving into uncharted waters. The role of system analyst as management consultant or management technical specialist comes into conflict with that of system analyst as

---

[1]System analysts work in a variety of arrangements including work as in-house staff and outside consultants. Their work also varies depending on the extent of interaction they have with users (i.e., products for software houses where users are often not directly involved in the development process, or user-specific applications in application areas like banks). The point here is that regardless of the type of work done, most developers are employed by management to carry out management goals. This chapter focuses on MIS applications, for discussion of product development organization see J. Grudin in this volume.

user-facilitator. For participatory design to work within the unstated assumptions of the American corporate world, these conflicting roles need to be spelled out and clarified. Movement from involving users in design to full participatory design means encroaching on some decisions that have traditionally been considered the prerogative of management. It also means looking at the concept of the integrated design of work organization, decision making, and computer support. In a sense participatory design opens up Pandora's box, for the questions we need to ask affect the whole organization and the way decisions are made within it.

To better understand this situation a brief look at the history of system development is useful. Following World War II some of the early methods and procedures for system development came from the field of Operations Research (OR). OR-based procedures emphasized quantitative reasoning in order to facilitate coding complex programs. In the 1950s as software for large mainframe computer systems was laboriously cranked out, practices from the newly emerging field of Management Science also began to frame the boundaries of system development field (see Greenbaum, 1979).

The merging of Operations Research and Management Science was not an accidental marriage. In the early days software development was extremely labor intensive and prone to error. System analysis and design emerged as a series of techniques that could place clear-cut standards on the way software was to be produced, including contractual arrangements to control the stages of development. Thus the system life cycle was defined as a series of linear steps from feasibility study through implementation and documentation, which could be seen by MIS departments as isolated project parts to be defined, administered and controlled. Interpreting this from the labor process perspective of Braverman's (1974) book on *Labor and Monopoly Capital,* one could say that computer systems were in fact designed to further divide workers and in many cases take away the skills that they had (Greenbaum, 1979, Kraft, 1977). Viewed in this way the system development process has not only excluded users from the development of systems, but has done so by design (Noble 1977).

The steps in the traditional life cycle approach were originally developed to control large scale projects yet the emphasis on *control* has remained a cornerstone of the profession. So much so, in fact, that if one were to read through the leading textbooks on System Analysis and Design today one would still find that the themes of dividing the project into stages and controlling each stage continue to dominate the material. Although newer textbooks introduce chapters on prototyping and inject some sections on the 'human factor', most material still focuses on procedures for meeting management needs for control over project costs time-frame requirements. In short, how the system might suit the needs of people working within the organization has been secondary to management requirements for control over the project.

As early as the 1960s some observers began to argue that the life cycle approach and formal procedures were unresponsive to human needs. In 1965,

Robert Boguslaw asserted that system developers were "concerned with neither souls nor stomachs," resulting in the fact that "People problems are left to the after-the-fact efforts of social scientists" (p.3). Today, close to 30 years after Boguslaw's warning, the tools and techniques taught to new system developers still follow the procedures for control outlined in the textbooks. In practice, the patterns of behavior among system developers and managers, for the most part, mirror the textbook methodology, or at least try to sound as if they do (see Friedman, 1989). System Analysis and Design as practiced within the confines of most Management Information Systems (MIS) departments looks a lot like general management practices with some specific techniques thrown in to account for the difficulty of controlling software projects.

Yet over the last 30 years the pages of management and system journals have been peppered with articles bemoaning the fact that so many systems don't work or fail to do things that both managers and users expect them to do. In addition to critiques within the system field, social scientists who have studied people at work point out that many computer systems adversely affect the performance of the workers using them. Most work places seem to have stories that confirm these findings—stories about workers having to do extra tasks just to make the data fit the system or about people spending long hours trying to print out letters on printers that are incompatible with their work stations.

General dissatisfaction about workplace systems seemed to have reached a feverish peak in the mid 1980s when MIS departments began to complain more loudly about taking the blame for systems that did not work (Friedman, 1989). Some changes that occurred during the last decade may help propel both MIS departments and users toward the need for some form of participatory design. The proliferation of PCs throughout many organizations means that many formerly passive users have begun to ask not what they could do for the MIS department, but what the MIS department could do for them. In fact, as many workers become more familiar with using computer hardware and software they no longer think of themselves as the silent majority. Thus the fact that MIS departments perceive that they are losing some control combined with the realization that users are getting more knowledgeable can lead us in the direction of practices that involve more participation from all who are affected by changes that occur when computer applications are installed or updated.

FROM USER-CENTERED TO PARTICIPATORY DESIGN

In the mid 1980s research in the area of user-centered design pointed out the need for applications that were not just user-friendly, but rather were more deeply rooted in the practices of people using them. The user-centered approach attempted to bring people back into the picture, putting emphasis on the need to develop systems that worked in practice, not just in testing. During this period studies from the field of human-computer interaction have played an important role in bringing the social sciences and humanities into the formerly quantitatively-oriented system development process (Norman & Draper, 1986). The focus on user-centered design has been important for both raising the issues of social interaction in the workplace and for developing new perspectives to frame the way systems, particularly human-computer interfaces, could be designed. Yet discussions about user-centered design arising mainly within product development organizations have not tackled the thornier issues of control in the workplace nor have they directly addressed decision-making in a MIS environment. The extent to which an American participatory design movement can openly and clearly take these next steps, will in my view, determine the success of building a bridge from user-centered to participatory activities.

As discussed in other chapters in this book, Scandinavian approaches to system development offer ideas and examples that could be applicable, with some modification, in North America. In particular, the approach called Cooperative Design places emphasis on cooperation between system developers and users and focuses on the cooperative nature of work. *Design at Work* (Greenbaum & Kyng, 1991) lays out an approach to cooperative computer system design. It outlines and develops a rationale for using techniques that fully involve users in the design process. These techniques stress the need for system developers to learn from the experiences of people using computers, not just from formal system descriptions of work, In addition, cooperative approaches argue that workplace language and daily experience of users need to be placed center stage in an effort to enable users. For enabling users implies not just using their experience, but creating and fostering an environment where they can feel empowered to express their ideas.

The background for the work on Cooperative Design grew out of both the need for more user-centered design and out of the Scandinavian worker-centered approaches to system development (Bjerknes, Ehn, & Kyng, 1987). While its origins lie in Scandinavia, Cooperative Design provides some necessary ingredients for U.S. system developers and users to begin the active and enabling process of working together. In this way Cooperative Design sets the stage for more participatory practices in both the United States and Scandinavia.

BUILDING PARTICIPATORY DESIGN IN THE UNITED STATES

I argue that the time is ripe for laying the groundwork for an American, "homegrown" participatory design movement. This does not mean that through design we could alter the power relations discussed in the second section of this chapter, but rather that we could encourage more active involvement of users and developers in the design process. Here are some of the events and situations which, I believe, we can use to build strategies that foster a Participatory Design environment.

Management Strategies—Team Work

Over the last several years management journals have focused on shifting managerial strategies, particularly those that emphasize the importance of team work. Economists have also addressed this issue in their discussions about the global challenges facing United States industry. While the challenges and changes reflect a broad spectrum of opinion, the majority of the discussions center around the ideas that team work is essential in virtually all jobs and that enhancing communication is a co-requisite for coordinating team work. The new managerial focus on team work and communication opens the door for developers and MIS departments to introduce ideas like those of Cooperative Design that include issues of work organization and computer support. For as managers worry about the ways that team work can be encouraged, participatory design techniques suggest concrete activities for applying these principles in the design of computer applications.

This is not to say that Participatory Design answers managements' problems, nor should it. Rather, it is a way for system developers to get in the door with strategies that indeed increase the likelihood of more worker participation, and offer concrete suggestions for designing systems that might better fit the working environment. Since most system projects start as a series of management objectives, it may now be more possible to make arguments that management is likely to hear. In particular, given the shift in management theory, system developers can argue that participatory design teams can play a role in fostering communications within organizations. These type of arguments do not, of course, solve the contradiction that system developers are caught in as they try to meet management objectives and worker needs. But, as the experiences outlined in this book indicate, Participatory Design could help create an environment where system developers and users of computer applications can learn to develop systems that better suit the way work is actually carried out. And it can be used as a wedge to develop frameworks that look at the whole context of work organization and build towards designing future work situations that reflect the needs of the people working in them.

Multicultural Work Groups

Another aspect of the current American experience grows out of discussions about the composition of the workforce and the importance of recognizing multicultural diversity. This movement, starting in the schools and slowly spreading to business organizations, stresses that we recognize that the workforce is made up of people from many cultures. Social scientists who study multiculturalism argue that this diversity can be used to enrich workplaces. On a very concrete level, multicultural pluralism lays out the arguments that support the need for participatory design. This movement recognizes that a workforce of people from many cultures not only speaks many languages, but may make differing assumptions about the nature of the work they are doing. The need to voice unspoken assumptions and make cultural pluralism a reality is a critical realization for American management. If taken seriously it could take the issues of civil rights to a new level, saying that civil rights in workplaces require acknowledging and respecting differences.

Obviously, the degree to which multicultural pluralism gains strength in the workplace will depend on political struggles that take place outside of work. At the same time that pluralism is being talked about in educational circles, courts in the United States have been moving away from worker rights. Yet as long as pluralism is being discussed, possibilities exist to convince management that techniques like those of Cooperative Design offer a way to bridge language and experience gaps. Emphasizing group process, experience, and workplace expertise, participatory processes could eat away at the old way of doing things. Specifically, projects that emphasize cooperation and group process are a way to shift the discussion away from some managers more narrow focus on control.

A PRAGMATIC APPROACH

Participatory design, as described in other chapters, involves process techniques. Some might argue that these process-oriented approaches may seem overly optimistic, yet they do provide a basis for system developers to make arguments that place their ideas on the list of possibilities that management is likely to consider. Experiences discussed in this book indicate ways that this is taking place. As Friedman (1989) points out, while some managers remain wedded to the issue of control, others have recognized that traditional system techniques fall far short of developing systems that work beyond their implementation phase. On a very basic level, many managers are ready to listen to new approaches because they are simply tired of taking the blame for bad systems.

Some advocates of participatory design have worried that the lack of a strong union movement would hamper its applicability in the United States. Certainly the high degree of unionization in Scandinavia, coupled with legislation that allows for worker discussions about technology, has helped build a base for participatory ideas in those countries (Greenbaum & Kyng, 1991). I believe that while it would be desirable to have that level of worker support, it is, within the current political environment, not a likely reality for American workers. Indeed by the end of the 1980s American union membership had fallen below 17% as traditional union strongholds in large industry lost jobs. Because white collar jobs, particularly Management Information Systems departments has never seen significant unionization, it would be useless to assume that union activity is a prerequisite to a participatory movement. Rather, I would argue that strong American opinions about the value of democracy and a long history of emphasizing process in both government and education could form a base for building participatory projects.

As in most other movements in American history, pragmatic approaches carry some weight. Participation of computer users in development projects offers pragmatic possibilities for management, users and system developers. For managers it offers the possibility of getting out of the hole of being blamed for projects that do not work; for users it clearly offers the opportunity to expand on their knowledge of the workplace and indeed to feel better integrated within it; and for system developers it offers the chance to build systems that work better. Like the civil rights movement and the women's movement, American activities may start with what seem to be rather idealistic goals, but once the economic roots of these ideas spread out, the possibilities for building on them grow larger. Of course there are set backs and side turns, but the process of involving more women and minorities in American workplaces has grown over the last 30 years. Similarly participation in technological decision making will take some time to take hold. The framework of beliefs is in place, and if introduced as both democratic and pragmatic, the ideas could spread out and grow.

The common ground for pragmatic workplace politics is clearly in place. The availability of 'off-the-shelf' packages and techniques for rapid prototyping provide evidence that the time frame of the traditional project life cycle can be shortened tremendously. Additionally, users are increasingly knowledgeable about computer applications, and certainly more vocal about their likes and dislikes. Even managers intent on maintaining tight control over project development have begun to see that the age of building 'idiot proof' systems may be coming to an end. Also more sophisticated software applications require more sophisticated users, making it in management's interest to work more closely with people who are using the new systems.

System developers can respond to these situations with a variety of arguments that open the door to increasing user participation. Borrowing from the Scandinavian approach, we could say that participation helps users increase their skills and thereby increase the quality of the services they provide (Bjerknes, et al., 1987). To managers bent on finding *solutions* that increase productivity, there is an obvious link between enhanced service and greater productivity. Indeed economists have found it almost impossible to measure increased productivity in the white collar sector, so rather than beating the same drum, managers could be encouraged to see that enhanced user participation and even the possibility of users providing better service might be more efficient than simply creating systems that increase paper or screen output.

Although to my knowledge, no cost studies of actual participatory design projects have yet been done, work-of-mouth reports indicate that involving users intensively and early in project development does not seem to increase project costs. And many argue that the increased time needed for these early group experiences, pays off in the likelihood that the project won't get bogged down toward the end when it is discovered that the system doesn't meet some crucial workplace need. Many MIS system developers have said that they had little trouble convincing managers that some regular release time was needed for users to get time off from regular duties in order to take part in design activities. As more and more of what gets done in companies begins to be recognized as part of participatory design, I believe that system developers will begin to swap success stories at an amazing rate. As noted in many conferences over the last few years, the stories of designing 'not by the book' have been a growing topic in hotel corridors and over coffee breaks.

I believe that the base exists to move from lip service to user participation to more active participation of users in the design process. More active participation, using the experiences discussed in this book, includes seeing users play an advisory role in decisions about computer support. And hopefully it includes watching users get a larger role in, at least advising, management about work organization and workplace environment issues. If we build on some of the experiences in this book we can clearly see that the *participatory* part of participatory design includes ways that: users gain more experience and knowledge about technology when they actively participate in project groups; system developers and users get better at designing and working with appropriate prototypes; users gain from getting hands-on experience of trying out possible software; and system developers do learn from the expertise of users—learning that in fact results in more workable prototypes and products.

FROM PARTICIPATION TO WORKPLACE DEMOCRACY

As discussed in the introduction to this chapter, the step from participatory design to workplace democracy is not automatic, for it involves serious shifts in control over decision making. The movement from active participation in an advisory capacity to actual decision making is a very big step. And the shift from system developers as MIS experts to more neutral facilitators is not so easily accomplished. Many system developers have found that while management might encourage them to actively enlist user participation that mandate does not include user decisions over the final system.

Americans are fed a steady diet of products and ideas that include the word *democracy*, yet within the confines of the workplace the concept is rarely if ever addressed. For at least one-third of every day, most Americans enter their work places and abandon discussion of this basic constitutional right. As in the discussion about management control over system projects, this lack of workplace democracy is no accident. For application of democratic ideas within the working environment would mean that on some level everyone could participate in decision making—decision-making that in the American framework is clearly management's prerogative.

This topic is broader than the subject matter in this book. Although my hope is that users as advisory participants can lead toward users as active participants in decision making, to date we have little to judge how this might happen in American workplaces. Certainly legislation of the sort that exists in Scandinavia would be necessary to help protect the ergonomic environment and to give workers the right to participate in decisions that affect their future workplaces— decisions that need to include the type of technology used and the applications for which it is used. The road from participatory design to workplace democracy can be built, in part, on the base of successful participatory design projects. This book and the experiences that people tell about the success of participatory approaches work toward building that road.

ACKNOWLEDGMENTS

This chapter grew out of a talk that I gave at the closing Plenary session of PDC'90. Lucy Suchman, as chair of the Plenary was inspirational in giving these ideas focus. The chapter is built mostly on my reflections about working with Cooperative Design when I was a guest at the Computer Science Department at Aarhus University, in Denmark, and in part on my experiences as a teacher and practitioner in the U.S. I thank the editors for lending me a copy of the tape of my talk and for helping me shape it into this chapter. Jonathan Grudin was wonderfully helpful in clarifying arguments and indicating muddled sentences. My Danish colleagues have also been helpful, as always, in pointing out the fuzzy parts of earlier drafts. Any remaining fuzzy ideas are, of course, my own.

REFERENCES

Bjerknes, G., Ehn, P., & Kyng, M. (Eds.). (1987). *Computers and democracy: A Scandinavian challenge.* Aldershot, UK: Avebury.

Boguslaw, R. (1965). *The new utopians—A study of system design and social change.* Englewood Cliffs, NJ: Prentice-Hall.

Braverman, H. (1974). *Labor and monopoly capital: The degradation of work in the twentieth century.* New York: Monthly Review Press.

Friedman, A. (1989). *Computer systems design: History, organiation and implementation.* Chicester, UK: Wiley.

Greenbaum, J. (1979). *In the name of efficiency, management theory, and shopfloor practice in data-processing work.* Philadelphia: Temple University Press.

Greenbaum, J., & Kyng, M. (Eds.). (1991). *Design at work: Cooperative design of computer systems.* Hillsdale, NJ: Lawrence Erlbaum Associates.

Kraft, P. (1977). *Programmers and managers: The routinization of computer programming in the U.S.* New York: Springer-Verlag.

Noble, D. (1977). *America by design: Science, technology and the rise of corporate capitalism.* New York: Knopf.

Norman, D., & Draper, S. (1986). *User centered system design: New perspectives on human-computer interaction.* Hillsdale, NJ: Lawrence Erlbaum Associates.

II PRINCIPLES AND ISSUES

4 Scandinavian Design: On Participation and Skill

Pelle Ehn
Aarhus University

In Scandinavia we have for two decades been concerned with participation and skill in the design and use of computer-based systems. Collaboration between researchers and trade unions on this theme, starting with the pioneering work of Kristen Nygaard and the Norwegian Metal Workers' Union, and including leading projects like DEMOS and UTOPIA, has been based on a strong commitment to the idea of industrial democracy. This kind of politically significant, interdisciplinary, and action-oriented research on resources and control in the processes of design and use has contributed to what is often viewed abroad as a distinctively Scandinavian approach to systems design.

This Scandinavian approach might be called a work-oriented design approach. Democratic participation and skill enhancement—not only productivity and product quality—themselves considered ends for the design. This chapter is based on two research projects, DEMOS and UTOPIA. I have elaborated this approach in detail in *Work-Oriented Design of Computer Artifacts* (1989).

Two important features of participatory design shape its trajectory as a design strategy. The political one is obvious. Participatory design raises questions of democracy, power, and control at the workplace. In this sense it is a deeply controversial issue, especially from a management point of view. The other major feature is technical—its promise that the participation of skilled users in the design process can contribute importantly to successful design and high-quality products. Some experiences, perhaps most developed in Scandinavia, support this prediction and contribute to the growing interest in participatory design in the United States and other countries; by contrast, "expert" design strategies have too often turned out to be failures in terms the usability of the resulting system. These two features together suggest that there is a firm link between the skill and product quality aspect of user participation and the democracy and control aspect, or else participatory design would be a deeply controversial issue from the point of view of the employees and trade unions.

The trade-union-oriented democracy aspect of skill and participation in design is discussed in the first part of this chapter. I start with an introduction to the concept of industrial democracy and an overview of the Scandinavian setting. After this background, research projects forming the Scandinavian work-oriented design approach are presented and discussed.

In the second part, I focus on the role of skill and participation in design as a practical activity. This focus has grown out of a dissatisfaction with traditional theories and methods for systems design. Not only has traditional design been oriented towards deskilling workers, but this traditional approach has been encouraged by a theoretical assumption, namely, that skill can be exhaustively characterized by a purely formal description. The political critique of the design process discussed in the first part leads to a theoretical critique of the scientific rationality of methods for systems description and systems development. In the second part, a philosophical foundation for a skill-based participatory design approach is outlined based on the language-game philosophy of Ludwig Wittgenstein. Taken together, these critiques shape the Scandinavian work-oriented design approach, an approach based on an emancipatory perspective and encompassing both the inner everyday life of skill-based participatory design and the societal and cultural conditions regulating this activity.

DESIGNING FOR DEMOCRACY AT WORK

The democratic ideal is a beautiful human invention: Every human should have the right to participate equally in decisions concerning his or her life. In practice this freedom has always been limited. The first democrats, the ancient Greeks, constrained participation to free men, excluding women and a class of slaves. The modern democratic state in capitalistic societies has, in theory and in many practical aspects, removed these constraints. Representative democracy is a formal arrangement for securing decisions in the interest of the majority, and often manages to assure freedom for minority groups.

In many sectors of life, however, democratic rights remain merely formal, without real content for those concerned. In this paper, I am concerned with democracy behind the factory gates and office doors—democracy at work.

Fundamentally, democracy at work or industrial democracy concerns freedom, another value-laden concept. It concerns *freedom from* the constraints imposed by the market economy and the power of capital. And it also concerns *freedom to* practically formulate and carry out particular projects that further democratize work. Attempts to democratize at work address:

1. The power of capital owners to control how resources are used. Possible uses include economic goals, structural changes in the company, investments in new technology, choice of business idea, and product range:
2. Owner's organizational and technological power to decide how the production process in general is organized and how technology is designed and used:
3. Their power over the workers to decide how work is to be organized, planned, and controlled:
4. Their power to limit an individual's autonomy at work, including the individual's choice of tools and pace of work (Dahlström, 1983).

The research projects on work-oriented design that I discuss in this paper address industrial democracy in all these aspects. They aimed both at a better understanding of freedom from managerial control and at freedom to develop and implement strategies for democratization at work. In particular, the studies looked at the design of computer-based systems in the context of democratization of work in Scandinavia.

The Scandinavian Setting

Scandinavian countries have for quite some time been well known for their distinctive industrial relations. The following features are particularly noteworthy:

1. A highly educated and relatively homogeneous workforce:
2. A high level of unionization:
3. Strong national trade union federations:
4. Centralized negotiation systems:
5. Large social democratic parties with strong links to the the national trade union federations of blue (and some white) collar workers, parties which for long periods of time have led the governments:
6. Relations between trade unions and employers that are, to a large extent, regulated by laws and central agreements:
7. A positive attitude to new technology from the trade union federations, at least since World War II and despite some opposition at the local level, based on the assumption that job loss due to the introduction of new technology would be compensated by active labor market government policies.

These features have contributed to the relative stability of Scandinavian labor relations and the relatively high degree of workplace democratization:

> Democracy [in the United States] stops at the office door and the factory gate. Western Europe is extending democracy into working life. Democratization of work has gone further in Scandinavia than elsewhere in Europe. Job redesign projects, codetermination arrangements, health and safety legislation, employee representatives on corporate boards. (Einhorn & Logue, 1982)

These historical factors help explain the emergence of the participatory trade-union-based work-oriented approach to design of computer artifacts. But just as important is the other side of the coin: The Scandinavian countries are still market economies, and an integral part of international capitalism. Workers and their unions therefore confront basically the same forces or rationalization of work and technology as those in other market economies.

Laws on Democratization of Work

The 1970s was the decade when democracy at work truly appeared on the agenda for industrial policy. In that decade, an intensive debate took place in trade unions, and a number of new labor laws were enacted (Fry, 1986).

In Norway, employees obtained the right to elect one-third of the members of the so-called "company assembly." In 1975, the first collective agreement on the development and introduction of computer-based systems was concluded, giving the trade unions the right to appoint so-called data shop stewards. In 1977, the Norwegian Work Environment Act gave workers extensive rights to stop production that was dangerous to their health. New codetermination procedures for work environment issues were established, and a system of sanctions was defined for employers who did not fulfil the new work environment requirements.

The Swedish "work democracy package" in the 1970s revised existing legislation and introduced several new acts. The work democracy package included the act concerning Labor Representatives on Company Boards, the Companies Act concerning disclosure of financial information, the act concerning the Status of Shop Stewards, and the Work Environment Act.

Finally, and most important, the Joint Regulation Act (MBL) concerning workers and trade unions' right to codetermination in production issues such as design and the use of new technology and work organization was enacted in January 1977. It was this law that the late Prime Minister Olof Palme described as the greatest reform in Swedish society since the introduction of the universal right to vote. In practice this act's impact has turned out to be far less dramatic, and as a result there was considerable disappointment among many union members who had received the impression that the Act was the decisive step towards democracy at work.

Nevertheless, MBL did create new conditions for the design and use of computer-based systems and other production equipment. Article 11 stipulates that the employer has to negotiate with the local union before making "major changes" in production. Article 12 give the union the right to initiate negotiations on any production issue. Articles 18 and 19 stipulate the right of unions to have access to documents to which management refers in negotiation, and to receive information continually on production issues, their employer's financial situation, and personnel policy.

These were important changes encouraging democratic control over the introduction and use of new technology, but they were limited. The Act gave the employer the exclusive right to make decisions when trade unions and management could not reach an agreement in negotiations. Furthermore, the "major changes" in production to which the Act referred are open to interpretation, as is the obligation to inform, which may or may not include early plans, say, to introduce a computer-based system.

Finally, Article 32 should be mentioned. This article concerns the right for trade unions to negotiate agreements on "the management and assignment of working duties, and the conduct of the operation at large." The number of this article was chosen to parallel Article 32 in the Swedish Employer's Confederation (SAFs) Statutes, which requires its members to retain the right of decision when entering collective agreements. The Act stipulates that if a collective codetermination agreement (MBA) is reached, the union has "priority of interpretation" over disputed issues covered by the agreement until the dispute is settled in negotiations. This gives the trade unions the opportunity to postpone decisions. However, the main idea behind Article 32 was that central agreements should be negotiated, and that local agreements should be developed on the basis of these agreements.

In 1978, the first central collective agreement on codetermination was reached in the public sector the (MBA-S). Not until 1982 was an agreement reached in private industry. By that time the forms of codetermination had become more concrete, but the trade unions' democratization offensive had to a significant extent faded out. What started as a trade union response to local demands for democratization in the late 1960s, often expressed as wild-cat strikes concerning the work environment and the introduction of new technology, had assumed a form sanctioned by parliament, national trade union federations, and national employers' federations.

The wild-cat strike by the workers' collective of iron miners at the Ikab mines in the north of Sweden was the starting point for these democratization reforms that concerned not only the democratization of the work place, but also internal trade union democracy. The work-oriented design approach emerged in the midst of the practical implementation of these reforms.

The Trade Unions as Vehicles for Industrial Democracy

In Scandinavia, trade unions have served as the vehicles for industrial democracy by advancing the interest of the workers' collective. The workers' collective is a concept developed by the Norwegian sociologist Sverre Lysgaard (1961) to designate the informal defense organization of workers in the workplace. The workers' collective is manifested as shared norms concerning how workers should behave in relation to management and the rationality of technical-economic organization. The norms shape workers' responses to management's efforts to intensify work, to tighten control of the labor process, and to rationalize the use of new technology. As hired labor, the workers are also part of the technical-economic organization of the workplace, but the workers' collective reflects workers' response to their subordinated position in this technical-economic organization.

According the Lysgaard, the degree of strength in the workers' collective comes from the "we-feeling" created by shared experiences. The basis for this we-feeling is physical proximity at the workplace, which makes interaction possible; similarity in working conditions, which makes the workers identify with each other; and a similar problem situation that they interpret in a similar way. The norms of the workers' collective define what it means to be a "good work mate" as well as what it means to be a "traitor." The workers' collective is a "buffer" between the individual worker and management's interests in shaping the technical-economic organization of the workplace.

However, the workers' collective has two major weaknesses when compared with institutionalized forms of industrial democracy. First, the workers' collective can only informally defend the individual worker. It is not an acknowledged formal organization, meaning that it has no formal organizational power for achieving structural changes in the workplace. Second, as a defensive organization responding to management initiative, it lacks the organizational ability to formulate and carry through an offensive strategy for changes in the direction of industrial democracy.

The trade union movement is a formal organization through which workers' interests can be developed and implemented. This by no means implies that trade unions always represent the interests of the workers' collective. Trade unions have a hierarchical structure, and the gap between shop-floor experiences and central decisions can be huge. Furthermore, the trade union movement is far from homogeneous. Different groups of workers have different interests. There are differences in interests and power between skilled and unskilled workers, between men and women, between workers organized in different trade unions, etc. In the design and introduction of new technology, these differences can manifest themselves as jurisdictional disputes and conflicts concerning skill development, work organization, and the right to operate the new technology. The problem of solidarity is a central one in the trade union movement's struggle for industrial democracy.

Nevertheless, trade unions have served the interest of the workers' collective in two ways that are important for the present argument: They have been an essential instrument for workers in wage negotiations, and they have been a key instrument for furthering democracy in society as a whole.

Unions and Design

The design and use of new technology requires new trade union activities. Traditionally, trade unions have focused on what Åke Sandberg (forthcoming) calls *distribution* issues such as wages, working hours, and general terms of employment. Such issues are characterized by:

1. Relatively well-developed union objectives;
2. Clearly formulated demands, often quantified;
3. Demands based on the workers' own practical experience; and
4. Clearly delimited, short negotiation cycles.

In contrast, the design and use of computer-based systems are, in Sandberg's terminology, *production* issues. They are characterized by:

1. Only vaguely formulated union objectives;
2. Demands that are difficult to quantify;
3. Practical on-the-job experiences that must be supplemented by more theoretical, technical/scientific knowledge;
4. Design processes that stretch over long periods of time; and
5. Negotiation situations that are difficult to define clearly.

As Sandberg points out, the design of new models for work seem to require the consideration of more deep-seated and qualitative aspects than can easily be fitted into the traditional trade union strategy or the traditional management-labor negotiation process.

What are the key elements of a union technology strategy? Obviously, one element is decentralization of decision making coupled with participation in the design process, which can give workers more influence and better access to important information. However, the position we took in DEMOS, UTOPIA, and the other projects that constituted the work-oriented design approach was that decentralization of decision making and a participative approach to the design process are not sufficient.

Our position is based on a recognition of the different interests of management and workers concerning industrial democracy. We rejected the harmony view of organizations, according to which conflicts in an organization are seen as stemming from misunderstandings and can resolved by good analysis. We also rejected an understanding of design as a rational decision making process based on common goals. Instead, our research was based on a conflict view of industrial organizations in our society. In the interest of emancipation, we deliberately made the choice of siding with workers and their organizations, supporting the development of their resources for a change towards democracy at work (Sandberg, 1979; Ehn & Sandberg, 1979). We found it necessary to identify with the we-feeling of the workers' collective rather than with the overall we-feeling and "modern management" attempts to create in order to elicit greater effort from the work force. Although trade unions had a structure that was problematic for functioning as vehicles for designing for democracy at work, they were also the only social force that in practice could be a carrier of that ideal.

From Sociotechnical Solutions to Work-Oriented Design

It is difficult to overstate the influence of the sociotechnical approach on user participation and industrial democracy initiatives (Kubicek, 1983) in Scandinavia in the early 1970s. Hence, it may appear paradoxical that some Scandinavian researchers and trade unions developed the work-oriented approach to democratization of design and use of computer-based systems in opposition to the sociotechnical tradition rather than within that tradition. I will try to explain why, and outline the main points of this alternative, "collective resource" approach.

Although initially implemented in Norway, the widespread use of the sociotechnical approach in Scandinavian industry took place in Sweden. Thoralf Qvale, in one of the evaluations of the sociotechnical approach to democratization in Norway, gives the following explanation:

> Apart from the researchers, there are very few persons trying to convey experience from one company to another. In Sweden, "job satisfaction" and "productivity" have been the slogans, and a network of employers/production engineers have taken care of the diffusion. In Norway, the slogans were "industrial democracy" and "participation," and the union networks were expected to play a central part. As explained, ideological support has come from the top of LO (the Norwegian national trade union federation), but the practical involvement from the individual unions' officials has systematically been lacking. (Qvale, 1976)

In the late 1960s in Sweden, rapid technological and structural change was considered a problem by both the trade unions and employers. The unions were concerned about deskilling, lack of influence, health, and safety. Employers experienced personnel problems in recruitment, turnover, and absenteeism, and production problems in efficiency, planning, and quality. Both parties came to see the Tayloristic organization of production—its narrowly specialized jobs and separation of conception and execution—as the source of these problems. The sociotechnical experiments in Norway seemed a promising way forward.

Similar experiments were initiated by the central unions and employer organizations jointly in both private and public sectors in Sweden. Increased job satisfaction and higher productivity were considered equally important goals in these tests. Several of the experiments that started in the late 1960s came up with interesting ideas on work organization and democratic participation, but practical implementation was a different story, as explained by Åke Sandberg:

> However, the second phase of deciding upon an actual program of change made manifest differences of interest: management was primarily seeking solutions to personnel problems and possibilities for better control of the wages, whereas unions viewed the experiments as part of a strategy for democratization and union influence at various levels. (Sandberg, 1982)

Most of the Swedish sociotechnical experiments were controlled by local management and coordinated by the Technical Department of the Swedish Employers' Confederation (SAF). The employers were obviously satisfied with the sociotechnical approach but not with the joint experiments. The LO (the Swedish national trade union federation) was also skeptical of joint work, as expressed in a program document:

> This method of working proved difficult to implement. Later, when the conflict of views between the two sides with regard to industrial democracy development became more manifest, the problems grew greater. Within the private sector, SAF drew its own conclusions from this fact and set up its own development projects with the aid of its Technical Department. In its development projects, SAF stressed the individual in a form which complicated collective solutions and the possibilities available to the trade union movement. (Swedish Federation of Trade Unions, 1977).

In 1975, SAF launched a new sociotechnical strategy. The Technical Department of SAF coordinated the "new factory" project, which aimed at creating more stable production systems based on the principle of coordinated independence of small subsystems (Arguren & Edgren, 1979). This principle was not new—it came from basic sociotechnical theory—but this project used a different strategy. On the one hand, it went further than earlier projects—it did not accept production technology as given, but as something that should be designed to allow the control of semiautonomous groups rather than individuals. On the other hand, changes were restricted to the shopfloor production level, and the vertical division of labor was not altered at all. Management's overall control was therefore strengthened. Democratic participation was not one of the aims of these experiments (Kronlund, 1978). The internationally known production technology at the Volvo Kalmar plant is as good an example as any of these new experiments. Participation for democracy was not an aim of the design (Ehn & Sandberg, 1979).

As we understand it, democratization of design and the use of computer-based systems in the Scandinavian setting had to be based on strong local union involvement. In practice, the sociotechnical approach had failed to support such democratization. The first action program on industrial democracy and computers from LO, issued in 1975, outlined the situation and union strategy:

> The workers and the trade unions are not satisfied with managers and their experts who say they develop systems for planning and control which take human beings into consideration by paying attention to needs for self-realization and the social impact of technical systems, etc. On the contrary, the unions must work for a situation that makes it possible for workers to develop their own organizational and knowledge resources. This creates the capability to scrutinize and influence, via negotiations, the various aspects of corporate planning and control, and, by extension, to develop worker-controlled systems. Thus, the present situation in organizations makes increasing demands upon the commitment and knowledge of the workers. The crucial point is whether these demands become absorbed in an employers' strategy for decentralization and so-called autonomous groups, or whether they will be developed within a workers' strategy for democratization, transcending the level of the work organization. (Swedish Federation of Trade Unions, 1975)

This critique of the sociotechnical approach should not be seen as a complete rejection of all aspects of the sociotechnical approach. Many of the sociotechnical tools are extremely useful in analyzing work organization and production technology, and the job requirements and group autonomy criteria are, when taken seriously, important criteria for democracy at work. The problem is that these requirements have often disappeared in the practical application of the approach. This flaw reflects questionable assumptions of harmony between social forces and a lack of sensitivity to the pervasive influence of the asymmetrical distribution of power. The critique is therefore not directed at sociotechnical design methods but at its theory and practice in the context of the democratization of work. We should also note that over the past decade the sociotechnical approach has evolved in a much more participatory and less manipulative direction (Gustavsen, 1985; Hedberg, 1980), making it more of an instrument for democratic design.

In the early 1970s, however, the sociotechnical approach seemed inadequate for democratizing the design and use of computer-based systems in the workplace. We had to look for an alternative based on a historical, social, and political understanding of the Scandinavian situation—an alternative that allowed the trade unions to play a major role. These were basic criteria in the emerging work-oriented approach to the design of computer-based systems.

The NJMF Project

In 1970, the Norwegian Iron and Metal Workers Union (NJMF) decided to initiate research of its own. When the NJMF project was first set up, the design was quite traditional. It involved a steering committee, a project group, and associated local unions at four different work places. The associated local unions were to act as reference groups. The project group consisted of two researchers and two staff members from the national union, and, according to the research plan, the researchers were to carry out a number of investigations in close cooperation with the two other members of the project group. Those investigations included:

1. A study of two or three computer-based planning and control systems;
2. A survey of the goals of the union in areas such as working conditions and control of organizations;
3. Formulation of demands on computer-based systems based on the survey;
4. An evaluation of the need for knowledge within NJMF in the areas of planning, control, and data processing, and possibly development of teaching material.

However, as the project progressed, it turned out to be impossible for the union people involved to apply the project's findings to the daily work at the factories, the local unions, and the national unions. The original project design had to a large extent been copied from a traditional research project approach used by managers and management consultants in a context where the goals were clear and the means for applying the project results had been discussed for decades. For the unions, there had been no extensive discussions on planning, control, and computer-based systems, and there were no established or clear goals for their involvement.

A completely new research strategy had to be developed. In the new strategy, the most important change was the new role to be played by the local unions. Instead of supporting the researchers, the researchers would support them. The local unions would choose the topics for study from important problems at the workplace, and they would receive assistance from external consultants as well as consultants and other resources provided by the company. At each of the four workplaces, a number of investigative groups consisting of union members was formed to

1. Accumulate knowledge about planning, control, and data processing;
2. Investigate selected problems in these areas, that were considered of special importance by the local unions; and
3. Take actions directed at management to change the use of new technology.

The groups always began with discussions of practical workplace problems, problems with which every worker was familiar. Attempts to analyze and to solve these problems led to the search for new knowledge and the start of an educational process. The groups met regularly, for two to three hours at least twice a week, and between the meetings, the members did a lot of "homework," such as preparing proposals, discussing ideas with fellow workers, and participating in different kinds of educational activities.

One investigation group made evaluations of some of the computer-based planning and control systems in the company, including an on-line production information system under development. The other investigation groups evaluated experiences of participation in the planning of a new plant, made proposals for reorganizing one of the main assembly lines, and drafted a company policy action program for the local union.

One of the most tangible, and certainly the most widely studied and publicized outcome, of the NJMF project was the earlier-mentioned data agreements. These agreements primarily regulate the design and introduction of computer-based systems, especially the availability of information. The first agreement, a local one, was signed at the beginning of 1974. It was followed in April 1975 by a central agreement between the Norwegian Trade Union Federation and the Norwegian Confederation of Employers. A large number of local agreements soon followed, as did the election of numerous so-called data shop stewards, a new kind of shop steward introduced in the central agreement.

Among other things, the central agreement stated:

> Through the shop stewards, the management must keep the employees orientated about matters that lie within the area of the agreement in such a way that the shop stewards can put forward their points of view as early as possible and before the management puts its decisions into effect. The orientation must be given in a well-arranged form and in a language that can be understood by nonspecialists. It is a condition that the representatives of the employees have the opportunity to make themselves acquainted with general questions concerning the influence of computer-based systems on matters that are of importance to the employees. The representatives must have access to all documentation about software and hardware within the area of the agreement. (Norwegian Employers Federation and Norwegian Federation of Trade Unions, 1975)

The NJMF project inspired several new research projects throughout Scandinavia and the development of a research tradition of cooperation between researchers and workers and their trade unions. This tradition is known as the *collective resource approach,* or the Scandinavian approach to work-oriented design.

The DEMOS Project

In Sweden, the DEMOS project on "trade unions, industrial democracy, and computers" started in 1975, and lasted four years. The Swedish Trade Union Confederation (LO) supported the project, with its "data council" acting as an advisory group. The project was carried out by an interdisciplinary research team (with competence in computer science, sociology, economics, and engineering) in cooperation with workers and their trade unions at four different enterprises— a daily newspaper, a locomotive repair shop, a metal factory, and a department store.

This cooperative effort tried to identify possibilities for the unions to influence the design and use of computer-based systems at the local level in the companies. It emphasized what the unions could do to safeguard and promote their members' interest in meaningful work when the technology, the work organization, and the supervision of work are altered. As a complement to these local activities, the project also sought to examine obstacles and limits confronting this democratization process.

The design work at a locomotive engine repair shop in Örebro serves as an example of the local approach to the DEMOS project. In 1974, the State Employees' Union was informed by the State Railway's central administration that a computer-based planning system, ISA-KLAR, would be introduced in its work shops in, among other places, Örebro. The main responsibility of the repair shop in Örebro was engine maintenance.

Management wanted to use ISA-KLAR to adapt the general maintenance system to local circumstances, and, in the process, to test ISA-KLAR. The union had won an elimination of piecework at the repair shop, and management hoped ISA-KLAR would help it redesign the workplace and develop automatic work orders and instructions to direct employees.

To implement ISA-KLAR, management formed project groups that included at least one trade union representative each. These groups interviewed workers in the workplace on how they carried out their jobs. The workers' tasks were then analyzed into smaller steps and the information coupled with an MTM database compiled from several big companies. ISA-KLAR used the combined information to specify detailed work steps, including their timing and sequence, and the tools to be used. The level of detail was very fine. The computer generated work sequences such as: (1) get tools A and B, (2) go to carriage, (3) crawl into position, (4) remove cotter pin, (5) remove washer and bolt, (6) repeat steps 3 through 5 for other bolts, (7) remove bolts, etc.

After two years' work with ISA-KLAR, the union turned to DEMOS to resolve dissatisfaction with its lack of influence and information on the project. In March 1976, the union established an investigative group of its own with 14 participants. At first, the local union asked for researchers from DEMOS to serve only as "data experts" to check the timings of the various tasks that were to be incorporated into the computer-based system for measuring performance. In subsequent discussions, the union researchers agreed that the key implication of ISA-KLAR—that work on the shopfloor could be deskilled—was a far more significant issue for investigation than the timing of steps in the performance of various tasks.

They shifted the investigation's focus to the whole question of production planning. There was further agreement that the union should conduct its own investigation, using the DEMOS researchers, separate from that of the management project groups. The DEMOS researchers would provide technical expertise in an analysis of the computer-based system and a structure for the union study. The union's basic investigation was completed in June 1977. Followup by the local union committee and researchers consumed another year.

The group's findings were summarized and transformed by the union into demands for local agreements of codetermination and rationalization. As a first step, the group issued a report describing the current work situation, pointing out what was good and what was bad. This description was then used as a basis for further studies regarding planning, control, and computer use, as well as for collective agreements within the Joint Regulation Act (MBL). It should be emphasized that the investigation was initiated before the introduction of MBL.

The report stressed that the changeover from piecework to a system of fixed monthly wages was extremely important, not only in terms of group solidarity among the workers and job satisfaction, but also in terms of the quality of production. Lack of planning and an uneven rate of work on the engines had created major problems, and the fact that neither the tools nor the spare parts were always available had been a great source of irritation at all workshops. Too much time was taken up searching for tools and spare parts. The poorly maintained work facilities had caused a number of problems in the work environment, such as draughts and working positions that were damaging workers' health. Uncertainty over the consequences of ISA-KLAR did not make the working conditions any better, and there was great dissatisfaction about the lack of information workers were receiving.

The investigation was completed by studies of various topical issues important to the trade union. These studies gave rise to a special group on planning, control, and computer use. This group compared the principles of Taylorism to the current work situation at the workshops and reviewed basic facts on computers, design, and planning methods in order to broaden worker discussions and to elicit views from as many members as possible.

One demand of the union—presented in various management project groups—was that problems with the planning, material administration, and work organization be solved before any discussion of computer-based time measurement. However, it became obvious that the union's chances of influencing the design of production planning by participating in the project groups were more or less illusory. Under the cover of technical discussions, management and its consultants continued to develop ISA-KLAR. The basic question of how planning in the shops would be conducted in the future never appeared on the agenda.

In response to management's investigation, and based on its own analysis of ISA-KLAR, the union's investigation group concluded that the system had to be stopped until an agreement on codetermination could be reached that would regulate its design and use. The local union demanded, and got, central union support for this position.

Management officially accepted the position of the union and appointed planning groups to "construct and test a planning model" for two different items in the production process (railway truck components and work on certain types of engines). Each of the planning groups consisted of two repair workers, a supervisor, and a production technician.

The investigation group of the union collaborated with the two planning groups. It appointed workers from its own investigation on ISA-KLAR as union participants and supported them. In practice the repair workers themselves did the design work. A technical specialist was asked to look over the proposals, and only minor adjustments were made.

Although the planning groups had very little time at their disposal, they managed to present concrete propositions on changes in work organization as well as other conditions. Their basic proposition was that repair workers should be granted flexibility in their work—predemolition, demolition, test of cracks, welding, installation, mounting of wheel axles, and final installation. Flexibility meant that all workers would participate in the entire work cycle, from demolition to final installation, and that all workers should all be able to handle all the tasks. The main emphasis were on skill, training, and job rotation. Special emphasis was placed on the work teams' right to plan their own work. This was considered necessary not only as a move towards democratic work organization but also as a measure facilitating production.

The repair workers claimed that their proposition had demonstrated that they could have a well-functioning workplace without ISA-KLAR. The local union that their approach could serve as a model for other workplaces as well. Its experiences were reflected in May 1978 in a number of demands that served as a basis for local agreements on rationalization and codetermination on the design and use of computer artifacts. The demands were adjusted to MBL and to a central collective agreement on codetermination. Among the demands were:

1. That long-term planning be conducted by management for (among other things) technical development, training, and staff policy;
2. That repair and maintenance work be carried out within the company;
3. That rationalization not reduce the requirements for skilled repair personnel; and
4. That rationalization not result in work measurement of individuals or groups or in incentive payments of any kind.

With special regard to the design process, the union demanded

1. That directives for a project be negotiable before the project starts;
2. That design methods be approved by the union;
3. That investigations in the design work not only include technical and economic considerations, but also changes in employment, work environment, work organization, and possibilities for cooperation, codetermination, and development in the daily work;
4. That the union be provided the necessary resources for conducting a parallel independent investigation;
5. That the cost for these resources be calculated as part of the investment in the rationalization;
6. That participation by trade union representatives and users be a natural aspect of the design work, and that it be planned to allow this; and
7. That participants receive what the union regards as the necessary training to participate in the design work.

What has been the long-term impact of the DEMOS investigation work at the repair shop in Örebro? We ourselves did not conduct such evaluations, but a report (Brulin, 1988) from an ongoing democratization project at the State Railroad sums up developments during the decade since DEMOS in the following way:

> The trade union work in Örebro from DEMOS onward has given a trade union perspective that perhaps is best formulated in the title of the DEMOS report: "We are opposed to Detailed Control." The collective agreement on design and use of computer-based systems created conditions for a dialogue with management. The repair shop in Örebro got a new managing director with a view on efficient organization that, to a great extent, overlaps the trade union perspective. In summary, as we understand it, what happened and is happening in Örebro can be explained by the above-mentioned three motivating forces: a trade union perspective, use of the collective agreement on design and use of computer-based systems, and a real dialogue with management.

The UTOPIA Project

Although growing, the extent and impact of research on designing for democratization in NJMF, DEMOS and similar projects did not meet initial expectations. It seemed that one could only influence the introduction of technology, training, and the organization of work to a certain degree. From a union perspective, important aspects for workers such as the opportunities to develop skills and to increase influence on work organization were limited. Societal constraints, especially those of power and resources, had been underestimated. In addition, the existing technology presented significant limits to finding desirable alternative local solutions.

To broaden the scope of available technology, we decided to try to supplement the existing elements of the work-oriented approach with union-based efforts to design new technology. The main idea of the first projects, to support democratization of the design process, was complemented by the idea of designing tools and environments for skilled work and good-quality products and services. To try out these ideas, the UTOPIA project was started in in 1981 as a cooperation between the Nordic Graphic Workers' Union and researchers in Sweden and Denmark with experience from the first generation of work-oriented design projects. It was a research project on the trade-union-based design of, and training in, computer technology and work organization. The research focused on page makeup and image processing in the newspaper industries. In the Scandinavian languages, UTOPIA is an acronym for Training, Technology, and Products from a Quality of Work Perspective.

Besides working directly in the project group, the Scandinavian graphic workers' unions followed and supported the project through a reference group consisting of representatives from Sweden, Denmark, Finland, and Norway, appointed by the Nordic Graphic Workers' Union (NGU). At various stages, the project involved the computer supplier Liber/Tips and the newspaper Aftonbladet in Stockholm.

In the UTOPIA project, we developed a design approach that we called the tool perspective (Ehn & Kyng, 1984; see also Bødker, forthcoming, and Kammersgaard, 1985). The tool perspective was deeply influenced by the way the design of tools takes place within traditional crafts. The idea is that new computer-based tools should be designed as an extension of the traditional practical understanding of tools and materials used within a given craft or profession. Design must therefore be carried out by the common efforts of skilled, experienced users and design professionals. Users possess the needed practical understanding but lack insight into new technical possibilities. The designer must understand the specific labor process that uses a tool. Computer-based tools present special challenges because they are technically complex but, if designed well, can be simple and powerful for the skilled worker.

In the UTOPIA project, we tried such a process of mutual learning. Graphics workers learned about the technical possibilities and constraints of computer technology, while we as designers learned about their craft or profession. Initially, the group worked to build a mutual understanding of the specific labor processes of the profession, of the design situation, and of the technical possibilities and limitations. Apart from discussions, visits to workplaces employing different generations of technology and visits to research laboratories and vendors proved to be important early activities.

However, as designers we ran into severe difficulties when we tried to communicate with the graphic workers using traditional approaches such as data or information flows. The situation drastically improved when we shifted towards a design-by-doing approach. With the use of mockups and other prototyping design artifacts, the skilled workers could actively participate in the design process and express their craft skills by actually doing page makeup. (The theoretical foundation of the tool perspective and these design methods will be discussed in the second part of this chapter.)

The UTOPIA project was in many ways a success story. An appreciative article in *Technology Review* concluded:

> So the impact of Utopia is continuing to expand, and the idea that workers and their unions have an important role in the design of new technology is reaching a wider and wider audience. Today Scandinavia, tomorrow, perhaps, the rest of the world. (Howard, 1985)

However, as in all success stories, UTOPIA had its share of failures as well. The failures were due to the limited resources in Scandinavian countries rather than to limitations of the model itself. The system, TIPS, was tried at several newspaper test sites, but before final development as a market product, the vendor ran short of capital and was forced to sell the rights to the system to another company interested mainly in image processing. These experiences indicate that a shortage of both technological competence and financial resources within small Scandinavian countries may keep them from successfully competing in the international technological race.

The UTOPIA project clearly showed that the latest technology may be designed and put into use to improve, not decrease, the skills of graphics workers. Whether the Scandinavian newspaper owners will exploit the possibilities for a constructive discussion on technology, organization, and training depends to a great extent upon whether the graphics workers and journalists succeed in overcoming their professional clash of interests and develop a common strategy.

The historical study conducted by the project provides some insight into this. New technology creates "demarcation disputes" between professional groups as well as between trade unions. The UTOPIA project demonstrated that solutions can, however, be found. For newspapers, there are technical and organizational alternatives that do not harm any professional group and that ensure product quality and reasonable efficiency. Nevertheless, the lack of trade union cooperation—rather than the technology, the newspaper owners, or the equipment vendors—may become the decisive factor frustrating the dream of UTOPIA.

The design process in the UTOPIA project was really utopian. The preconditions for such a design process are not present in corporate business as we know it today. Resources for skilled workers, trade union staff, and computer and social scientists to work together over a long period of time designing tools in the interest of the end users do not generally exist as yet, not even in Scandinavia,

UTOPIA was not only a challenge to design, but also to a more democratic working life.

Other Projects

The NJMF, DEMOS, and UTOPIA projects are by no means the only projects of the participatory work-oriented design approach, nor is the approach restricted to the design and use of computer-based systems. Some other projects within or related to this Scandinavian tradition are:

1. The DUE project on "democracy, education, and computer-based systems" was a sister project to DEMOS carried out in Denmark (Kyng & Matthiassen, 1982).
2. The Dairy Project, which was conducted by architects but used methods and perspectives similar to the DEMOS and DUE projects (Steen & Ullmark, 1982).
3. The PAAS project, which in addition to contributing to a theoretical understanding of changes of skills when computer artifacts are used also contributed to methods for trade union design work (Göranzon, 1984).
4. The Bank Project, which was conducted by researchers originally from the sociotechnical tradition, although they worked closely with trade unions and with methods and perspectives very similar to the work-oriented design projects (Hedberg & Mehlmann, 1983).
5. The TIK-TAK project, in which local trade unions in the public sector developed union resources in relation to "office automation" (Foged et al., 1987).
6. The Carpentry Shop project, which worked with methods and a design perspective similar to the UTOPIA project but within a "low-tech" area (Sjögren, 1979–83).
7. The Florence project, focusing on the work situation of nurses, which was another second-generation work-oriented project designing computer-based environments for skill and quality production (Bjerknes & Bratteteig, 1987).

These are by no means the only work-oriented design projects. Today the approach is no longer limited to Scandinavia. Despite a very different trade union structure, there are several projects using similar perspectives and methods in Britain (Williams, 1987).

Some Lessons on Design and Democratization

Some of the *general lessons* learned from DEMOS, UTOPIA, and other work-oriented design projects include:

1. A participatory approach to the design process is not sufficient in the context of democratization at work.
2. In democratization of design and use of computer-based systems in Scandinavia, trade unions—especially on a local level—must play an active role.

Some specific lessons about the *participation of local trade unions* in the design of computer-based systems include:

1. A clear distinction based on negotiations between union and management roles in the design process is not in opposition to, but a prerequisite for cooperation and the democratization of decision making in the work organization.
2. The design and use of computer-based systems requires new trade union activities.
3. The most important prerequisite for trade union participation in the design process is a parallel and independent process of accumulation of knowledge on the part of the union.
4. Local unions need external resources and support in their design activities.
5. A local trade union strategy has to be based on solidarity between the different groups of workers involved—a solidarity that goes beyond the traditional division of labor in the labor process and the traditional jurisdictions between the unions involved.

Some specific lessons on *national trade union support for democratization* of the design and use of computer-based systems include:

1. Today's computer-based systems often restrict the ability of trade unions to reach local objectives, especially with respect to skill but also with respect to work organization.
2. However, it is possible to design computer-based technology based on criteria such as skill and democracy at work.
3. National trade unions must influence the process of research and development of new technology to change the supply of technological and organizational solutions.
4. Equally important is a trade union strategy to influence the demand for these technological and organizational alternatives.
5. National trade unions must provide training with a trade union perspective on the design and use of computer artifacts, and influence the supply of professional training for skilled work.
6. A strategy like the new Scandinavian model for research and technological

development—focusing on a new form of cooperation among governments, trade unions, and high tech industry in the production of new technology that supports good working conditions and good use quality products and services—is a promising approach to support more democratic design and use of computer-based systems.

TOWARDS A PHILOSOPHICAL FOUNDATION FOR SKILL-BASED PARTICIPATORY DESIGN

This paper does not argue for a reinvention of the wheel. The instrumental power of systems thinking for purposive rational action is beyond doubt. Many of the computer applications that function well today could not have been designed without rational methods. Instead, I suggest a reinterpretation of design methods to take us beyond the deeply embedded Cartesian mind-body dualism and beyond the limits of formalization towards an understanding that supports more creative ways of thinking and doing design as participatory work (involving the skills of both users and designers).

Efforts to pursue such a rethinking of the design of computer-based systems and to develop a new practice of design are now emerging within computer science. One important example is a new orientation in software engineering proposed by Christiane Floyd (1987). It is based on a dissatisfaction with "anomalies" in the product-oriented view of software engineering that treats computer programs as formal mathematical objects derived by formalized procedures from an abstract specification. Floyd argues that the product-oriented view leaves the relationship between programs and the living human world entirely unexplored, providing no way to check the relevance of the specification or to accommodate learning and communication.

As a remedy to these anomalies, Floyd sees a new process-oriented paradigm in software engineering with a focus on human learning and communication in both the use and development of the software. She views the products of this process as tools or working environments for people and not as pieces code or an abstract software system. Hence, the quality of the product depends on its relevance, suitability, or adequacy in practical use. Quality cannot be reduced to features of the product such as reliability and efficiency. From this perspective, prototyping can be seen as an alternative or complement to traditional, more formalized, and detached descriptions.

Another important example of new tendencies in the design of computer-based systems is the development of a new philosophical foundation in the tradition of hermeneutics and phenomenology proposed by Hubert and Stuart Dreyfus (1986) and Terry Winograd and Fernando Flores (1986). This philosophical endeavor focuses on the differences between human activity and computer performance. In doing so, it departs from other traditions by focusing on what people *do* with computers, how in cooperation with one another they use computers, and what they might do better with computers. In this approach, the origin of design is in involved practical use and understanding, not detached reflection, and design is seen as an interaction between understanding and creation. This research aims not to create just another design method but to create a new foundation for a science of design.

In the following, I will propose that this new understanding can be buttressed by an awareness of language games and the ordinary language philosophy of Ludwig Wittgenstein. My focus is on the shift in design from *language as description* towards *language as action*.

Rethinking Systems Descriptions

A few years ago I was struck by something I had not noticed before. While thinking about how perspectives make us select certain aspects of reality as important in a description, I realized I had completely overlooked my own presumption that descriptions in one way or another are mirror images of a given reality. My earlier reasoning had been that because there are different interests in the world, we should always question the objectivity of design choices that claimed to flow from design as a process of rational decision making. Hence, I had argued that we needed to create descriptions from different perspectives in order to form a truer picture. I did not, however, question the Cartesian epistemology and ontology of an inner world of experiences (mind) and an outer world of objects (external reality). Nor did I question the assumption that language was our way of mirroring this outer world of real objects. By focusing on which objects and which relations should be represented in a systems description, I took for granted the Cartesian mind-body dualism that Wittgenstein had so convincingly rejected in *Philosophical Investigations* (1953). Hence, although my purpose was the opposite, my perspective blinded me to the subjectivity of craft, artistry, passion, love, and care in the system descriptions.

Our experiences with the UTOPIA project caused me to re-examine my philosophical assumptions. Working with the end users of the design, the graphics workers, some design methods failed while others succeeded. Requirement specifications and systems descriptions based on information from interviews were not very successful. Improvements came when we made joint visits to interesting plants, trade shows, and vendors and had discussions with other users; when we dedicated considerably more time to learning from each other, designers from graphics workers and graphics workers from designers; when we started to use design-by-doing methods and descriptions such as mockups and work organization games; and when we started to understand and use traditional tools as a design ideal for computer-based systems.

The turnaround can be understood in the light of two Wittgensteinian lessons. The first is not to underestimate the importance of skill in design. As Peter Winch (1958) has put it, "A cook is not a man who first has a vision of a pie and then tries to make it. He is a man skilled in cookery, and both his projects and his achievements spring from that skill." The second is not to mistake the role of description methods in design: Wittgenstein argues convincingly that what a picture describes is determined by its use.

In the following I will illustrate how our "new." UTOPIAN design methods may be understood from a Wittgensteinian position, that is, why design-by-doing and a skill-based participatory design process works. More generally, I will argue that design tools such as models, prototypes, mockups, descriptions, and representations act as reminders and paradigm cases for our contemplation of future computer-based systems and their use. Such design tools are effective because they recall earlier experiences to mind. It is in this sense that we should understand them as *representations*. I will begin with a few words on practice, the alternative to the "picture theory of reality".

Practice is Reality

Practice as the social construction of reality is a strong candidate for replacing the picture theory of reality. In short, practice is our everyday practical activity. It is the human form of life. It precedes subject-object relations. Through practice, we produce the world, both the world of objects and our knowledge about this world. Practice is both action and reflection. But practice is also a social activity; it is produced in cooperation with others. To share practice is also to share an understanding of the world with others. However, this production of the world and our understanding of it takes place in an already existing world. The world is also the product of former practice. Hence, as part of practice, knowledge has to be understood socially—as producing or reproducing social processes and structures as well as being the product of them (Kosik, 1967; Berger & Luckmann, 1966).

Against this background, we can understand the design of computer applications as a concerned social- and historical-conditioned activity in which tools and their use are envisioned. This is an activity and form of knowledge that is both planned and creative.

Once struck by the "naive" Cartesian presumptions of a picture theory, what can be gained in design by shifting focus from the correctness of descriptions to intervention into practice? What does it imply to take the position that what a picture describes is determined by its use? Most importantly, it sensitizes us to the crucial role of skill and participation in design, and to the opportunity in practical design to transcend some of the limits of formalization through the use of more action-oriented design artifacts.

Language as Action

Think of the classical example of a carpenter and his or her hammering activity. In the professional language of carpenters, there are not only hammers and nails. If the carpenter were making a chair, other tools used would include a drawknife, a brace, a trying plane, a hollow plane, a round plane, a bow-saw, a marking gauge, and chisels (Seymour, 1984). The materials that he works with are elm planks for the seats, ash for the arms, and oak for the legs. He is involved in saddling, making spindles, and steaming.

Are we as designers of new tools for chairmaking helped by this labeling of tools, materials, and activities? In a Wittgensteinian approach the answer would be; only if we understand the practice in which these names make sense. To label our experiences is to act deliberately. To label deliberately, we have to be trained to do so. Hence, the activity of labeling has to be learned. Language is not private but social. The labels we create are part of a practice that constitutes social meaning. We cannot learn without learning something specific. To understand and to be able to use is one and the same (Wittgenstein, 1953). Understanding the professional language of chairmaking, and any other language-game (to use Wittgenstein's term), is to be able to master practical rules we did not create ourselves. The rules are techniques and conventions for chairmaking that are an inseparable part of a given practice.

To master the professional language of chairmaking means to be able to act in an effective way together with other people who know chairmaking. To "know" does not mean explicitly knowing the rules you have learned, but rather recognizing when something is done in a correct or incorrect way. To have a concept is to have learned to follow rules as part of a given practice. Speech acts are, as a unity of language and action, part of practice. They are not descriptions but actions among others in a given practice.

Below I will elaborate on language-games, focusing on the design process, descriptions in design, design artifacts, and knowledge in the design of computer applications.

Language Games

To use language is to participate in language-games. In discussing how we in practice follow (and sometimes break) rules as a social activity, Wittgenstein asks us to think of games, how they are made up and played. We often think of games in terms of a playful, pleasurable engagement. I think this aspect should not be denied, but a more important aspect for our purpose here is that games are mainly interested activities, as are most of the common language-games we play in our ordinary language.

Language-games, like the games we play as children, are social activities. To be able to play these games, we have to learn to follow rules, rules that are socially created but far from always explicit. The rule-following behavior of being able to play together with others is more important to a game than the specific explicit rules. Playing is interaction and cooperation. To follow the rules in practice means to be able to act in a way that others in the game can understand. These rules are embedded in a given practice from which they cannot be distinguished. To know them is to be able to "embody" them, to be able to apply them to an open class of cases.

We understand what counts as a game not because we have an explicit definition but because we are already familiar with other games. There is a kind of family resemblance between games. Similarly, professional language-games can be learned and understood because of their family resemblance to other language-games that we know how to play.

Language games are performed both as speech acts and as other activities, as meaningful practice within societal and cultural institutional frameworks. To be able to participate in the practice of a specific language-game, one has to share the form of life within which that practice is possible. This form of life includes our natural history as well as the social institutions and traditions into which we are born. This condition precedes agreed social conventions and rational reasoning. Language as a means of communication requires agreement not only in definitions, but also in judgments. Hence, intersubjective consensus is more fundamentally a question of shared background and language than of stated opinions (Wittgenstein, 1953).

This definition seems to make us prisoners of language and tradition, which is not really the case. Being socially created, the rules of language games, like those of other games, can also be socially altered. There are, according to Wittgenstein, even games in which we make up and alter the rules as we go along. Think of systems design and use as language games. The very idea of the interventionistic design language-game is to change the rules of the language-game of use in a proper way.

The idea of language-games entails an emphasis on how we linguistically discover and construct our world. However, language is understood as our use of it, as our social, historic, and intersubjective application of linguistic artifacts. As I see it, the language-game perspective therefore does not preclude consideration of how we also come to understand the world by use of other tools.

Tools and objects play a fundamental role in many language-games. A hammer is in itself a sign of what one can do with it in a certain language-games. And so is a computer application. These signs remind one of what can be done with them. In this light, an important aspect in the design of computer applications is that its signs remind the users of what they can do with the application in the language-games of use (Brock, 1986). The success of "what-you-see-is-what-you-get" and "direct manipulation" user interfaces does not have to do with how they mirror reality in a more natural way, but with how they provide better reminders of the users' earlier experiences (Bødker, forthcoming). This is also, as will be discussed in the following, the case with the tools that we use in the design process.

Knowledge and Design Artifacts

As designers we are involved in reforming practice, in our case typically computer-based systems and the way people use them. Hence, the language-games of design change the rules for other language-games, in particular those of the application's use. What are the conditions for this interplay and change to operate effectively?

A common assumption behind most design approaches seems to be that the users must be able to give complete and explicit descriptions of their demands. Hence, the emphasis is on methods to support this elucidation by means of requirement specifications or system descriptions (Jackson, 1983; Yourdon, 1982).

In a Wittgensteinian approach, the focus is not on the "correctness" of systems descriptions in design, on how well they mirror the desires in the mind of the users, or on how correctly they describe existing and future systems and their use. Systems descriptions are design artifacts. In a Wittgensteinian approach, the crucial question is how we use them, that is, what role they play in the design process.

The rejection of an emphasis on the "correctness" of descriptions is especially important. In this, we are advised by the author of perhaps the strongest arguments for a picture theory and the Cartesian approach to design—the young Wittgenstein in *Tractatus Logico-Philosophicus* (1923). The reason for this rejection is the fundamental role of practical knowledge and creative rule following in language-games.

Nevertheless, we know that systems descriptions are useful in the language-game of design. The new orientation suggested in a Wittgensteinian approach is that we see such descriptions as a special kind of artifact that we use as "typical examples" or "paradigm cases." They are not models in the sense of Cartesian mirror images of reality (Nordenstam, 1984). In the language-game of design, we use these tools as reminders for our reflection on future computer applications and their use. By using such design artifacts, we bring earlier experiences to mind, and they bend our way of thinking of the past and the future. I think that this is why we should understand them as *re*presentations (Kaasbøll, forthcoming). And this is how they inform our practice. If they are good design artifacts, they will support good moves within a specific design language-game.

The meaning of a design artifact is its use in a design language-game, not how it "mirrors reality." Its ability to support such use depends on the kinds of experience it evokes, its family resemblance to tools that the participants use in their everyday work activity. Therein lies a clue to why the breakthrough in the UTOPIA project was related to the use of prototypes and mockups. Since the design artifacts took the form of reminders or paradigm cases, they did not merely attempt to mirror a given or future practice linguistically. They could be experienced through the practical use of a prototype or mockup. This experience could be further reflected upon in the language-game of design, either in ordinary language or in an artificial one.

A good example from the UTOPIA project is an empty cardboard box with "desktop laser printer" written on the top. There is no functionality in this mockup. Still, it works very well in the design game of envisioning the future work of makeup staff. It reminded the participating typographers of the old "proof machine" they used to work with in lead technology. At the same time, it suggested that with the help of new technology, the old proof machine could be reinvented and enhanced.

This design language-game was played in 1982. At that time, desktop laser printers only existed in advanced research laboratories, and certainly typographers had never heard of them. To them, the idea of a cheap laser printer was "unreal."

It was our responsibility as professional designers to be aware of such future possibilities and to suggest them to the users. It was also our role to suggest this technical and organizational solution in such a way that the users could experience and envision what it would mean in their practical work, before the investment of too much time, money, and development work. Hence, the design game with the mockup laser printer. The mockup made sense to all participants—users and designers (Ehn & Kyng, 1991).

This focus on nonlinguistic design artifacts is not a rejection of the importance of linguistic ones. Understood as triggers for our imagination rather than as mirror images of reality, they may well be our most wonderful human inventions. Linguistic design artifacts are very effective when they challenge us to tell stories that make sense to all participants.

Practical Understanding and Propositional Knowledge

There are many actions in a language-game, not least in the use of prototypes and mockups, that cannot be explicitly described in a formal language. What is it that the users know, that is, what have they learned that they can express in action, but not state explicitly in language? Wittgenstein (1953) asks us to "compare knowing and saying: how many feet high Mont Blanc is—how the word 'game' is used—how a clarinet sounds. If you are surprised that one can know something and not be able to say it, you are perhaps thinking of a case like the first. Certainly not of one of the third."

In the UTOPIA project, we were designing new computer applications to be used in typographical page makeup. The typographers could tell us the names of the different tools and materials that they use such as knife, page ground, body text, galley, logo, halftone, frame, and spread. They could also tell when, and perhaps in which order, they use specific tools and materials to place an article. For example, they could say, "First you pick up the body text with the knife and place it at the bottom of the designated area on the page ground. Then you adjust it to the galley line. When the body text fits you get the headline, if there is not a picture," and so forth. What I, as designer, get to know from such an account is equivalent to knowing the height of Mont Blanc. What I get to know is very different from the practical understanding of really making up pages, just as knowing the height of Mont Blanc gives me very little of understanding the practical experience of climbing the mountain.

Knowledge of the first kind has been called *propositional knowledge*. It is what you have "when you know that something is the case and when you also can describe what you know in so many words" (Nordenstam, 1985). Propositional knowledge is not necessarily more reflective than practical understanding. It might just be something that I have been told, but of which I have neither practical experience nor theoretical understanding.

The second case, corresponding to knowing how the word *game* is used, was more complicated for our typographers. How could they, for example, tell us the skill they possess in knowing how to handle the knife when making up the page in pasteup technology? This is their practical experience from the language-games of typographic design. To show it, they have to do it.

And how should they relate what counts as good layout, the complex interplay of presence and absence, light and dark, symmetry and asymmetry, uniformity and variety? Could they do it in any other way than by giving examples of good and bad layouts, examples that they have learned by participating in the games of typographical design? As in the case of knowing how a clarinet sounds, this is typically sensuous knowing by familiarity with earlier cases of how something is, sounds, smells, and so on.

Practical understanding—in the sense of practical experience from doing something and having sensuous experiences from earlier cases—defies formal description. If it were transformed into propositional knowledge, it would become something totally different.

It is hard to see how we as designers of computer systems for page makeup could manage to come up with useful designs without understanding how the knife is used or what counts as good layout. For this reason we had to have access to more than what can be stated as explicit propositional knowledge. We could only achieve this understanding by participating to some extent in the language-games of use of the typographical tools. Hence, participation applies not only to users participating in the language-game of design, but perhaps more importantly to designers participating in use. Some consequences of this position for organizing design language-games will be discussed in the following.

Rule Following and Tradition

Now, I turn to the paradox of rule-following behavior. As mentioned, many rules that we follow in practice can scarcely to be distinguished from the behavior in which we perform them. We do not know that we have followed a rule until we have done it. The most important rules we follow in skillful performance defy formalization, but we still understand them.

As Michael Polanyi (1973), the philosopher of tacit knowledge, has put it: "It is pathetic to watch the endless efforts—equipped with microscopy and chemistry, with mathematics and electronics—to reproduce a single violin of the kind the half-literate Stradevarius turned out as a matter of routine more than 200 years ago." This is the traditional aspect of human rule-following behavior. Polanyi points out that what may be our most widely recognized, explicit, rule-based system—the practice of Common Law—also uses earlier examples as paradigm cases. Says Polanyi, "[Common Law] recognizes the principle of all traditionalism that practical wisdom is more truly embodied in action than expressed in the rules of action." According to Polanyi this is also true for science, no matter how rationalistic and explicit it claims to be: "While the articulate contents of science are successfully taught all over the world in hundreds of new universities, the unspecifiable art of scientific research has not yet penetrated to many of these." The art of scientific research defies complete formalization; it must be learned partly by examples from a master whose behavior the student trusts.

Involving skilled users in the design of new computer application when their old tools and working habits are redesigned is an excellent illustration of Polanyi's thesis. If activities that have been under such pressure for formalization as Law and Science are so dependent on practical experience and paradigm cases, why should we expect other social institutions that have been under less pressure of formalization to be less based on practical experience, paradigm cases, and tacit knowledge?

Rule Following and Transcendence

If design is rule-following behavior, is it also creative transcendence of traditional behavior. Again, this is what is typical of skillful human behavior, and is exactly what defies precise formalization. Through mastery of the rules comes the freedom to extend them. This creativity is based on the open-textured character of rule-following behavior. To begin with, we learn to follow a rule as a kind of dressage, but in the end we do it as creative activity (Dreyfus & Dreyfus, 1986). Mastery of the rules puts us in a position to invent new ways of proceeding. As the Wittgenstein commentator Alan Janik has put it: "There is always and ineliminably the possibility that we can follow the rule in a wholly unforeseen way. This could not happen if we had to have an explicit rule to go on from the start . . . the possibility of radical innovation is, however, the logical limit of description. This is what tacit knowledge is all about" (Janik, 1988). This is why we need a strong focus on skill both in design and in the use of computer systems. We focus on existing skills, not at to inhibit creative transcendence, but as a necessary condition for it.

But what is the role of "new" external ideas and experiences in design? How are tradition and transcendence united in a Wittgensteinian approach? It could, I believe, mean utilizing something like Berthold Brecht's theatrical "alienation" effect *Verfremdungseffekt* to highlight transcendental untried possibilities in everyday practice by presenting a well-known practice in a new light: "the aspects of things that are most important to us are hidden because of their simplicity and familiarity" (Wittgenstein, 1953). However, as Peter Winch (1958, p. 119) put it, in a Wittgensteinian approach: "the only legitimate use of such a *Verfremdungseffekt* is to draw attention to the familiar and obvious, not to show that it is dispensable from our understanding."

Design artifacts, linguistic or not, may in a Wittgensteinian approach certainly be used to break down traditional understanding, but they must make sense in the users' ordinary language-games. If the design tools are effective, it is because they help users and designers to see new aspects of an already well-known practice, not because they convey such new ideas. It is I think fair to say that this focus on traditional skill in interplay with design skill may be a hindrance to really revolutionary designs. The development of radically new designs might require leveraging *other* skills and involving *other* potential users. Few designs, however, are really revolutionary, and for normal everyday design situations, the participation of traditionally skilled users is critical to the quality of the resulting product.

The tension between tradition and transcendence is fundamental to design. There can be a focus on tradition or transcendence in the systems being created. Should a word processor be designed as an extension of the traditional typewriter or as something totally new? Another dimension is professional competence: Should one design for the "old" skills of typographers or should new knowledge replace those skills in future use? Or again, with the division of labor and cooperation: Should the new design support the traditional organization in a composing room or suggest new ways of cooperation between typographers and journalists? There is also the tension between tradition and transcendence in the goods or services to be produced using the new system: Should the design support the traditional graphical production or completely new services, such as desktop publishing?

Tradition and transcendence, that is the dialectical foundation of design.

Design by Doing: New "Rules of the Game"

What do we as designers have to do to qualify as participants in the language-games of the users? What do users have to learn to qualify as participants in the language-game of design? And what means can we develop in design to facilitate these learning processes?

If designers and users share the same form of life, it should be possible to overcome the gap between the different language-games. It should, at least in principle, be possible to develop the practice of design to the point where there is enough family resemblance between a specific language-game of the users and the language-games in which the designers of the computer application are intervening. A mediation should be possible.

But what are the conditions required to establish this mediation? For Wittgenstein, it would make no sense to ask this question outside a given form of life: "If a lion could talk, we could not understand him" (1953). In the arguments below, I have assumed that the conditions for a common form of life are possible to create, that the lions and sheep of industrial life, as discussed in the first part of this chapter, can live together. This is more a normative standpoint of how design ought to be, a democratic hope rather than a reflection on current political conditions.

To develop the competence required to participate in a language-game requires a lot of learning within that practice. But, in the beginning, all one can understand is what one has already understood in another language-game. If we understand anything at all, it is because of the family resemblance between the two language-games.

What kind of design tools could support this interplay between language-games? I think that what we in the UTOPIA project called design-by-doing methods—prototyping, mockups, and scenarios—are good candidates. Even joint visits to workplaces, especially ones similar to the ones being designed for, served as a kind of design tool through which designers and users bridged their language-games.

The language-games played in design-by-doing can be viewed both from the point of view of the users and of the designers. This kind of design becomes a language-game in which the users learn about possibilities and constraints of new computer tools that may become part of their ordinary language-games. The designers become the teachers that teach the users how to participate in this particular language-game of design. However, to set up these kind of language-games, the designers have to learn from the users.

However, paradoxical as it sounds, users and designers do not have to understand each other fully in playing language-games of design-by-doing together. Participation in a language-game of design and the use of design artifacts can make constructive but different sense to users and designers. Wittgenstein (1953) notes that "when children play at trains their game is connected with their knowledge of trains. It would nevertheless be possible for the children of a tribe unacquainted with trains to learn this game from others, and to play it without knowing that it was copied from anything. One might say that the game did not make the same sense as to us." As long as the language-game of design is not a nonsense activity to any participant but a shared activity for better understanding and good design, mutual understanding may be desired but not really required.

User Participation and Skill

The users can participate in the language-game of design because the application of the design artifacts gives their design activities a family resemblance with the language-games that they play in ordinary use situations. An example from the

UTOPIA project is a typographer sitting at a mockup of a future workstation for page makeup, doing page makeup on the simulated future computer tool.

The family resemblance is only one aspect of the methods. Another aspect involves what can be expressed. In design-by-doing, the user is able to express both propositional knowledge and practical understanding. Not only could, for example, the typographer working at the mockup tell that the screen should be bigger to show a full page spread—something important in page makeup—he could also show what he meant by "cropping a picture" by actually doing it as he said it. It was thus possible for him to express his practical understanding, his sensuous knowledge by familiarity. He could, while working at the mockup, express the fact that when the system is designed one way he can get a good balanced page, but not when it is designed another way.

Designer Participation and Skill

For us as designers, it was possible to express both propositional knowledge and practical understanding about design and computer systems. Not only could we express propositional knowledge such as "design-by-doing design tools have many advantages as compared with traditional systems descriptions" or "bit-map displays bigger than 22 inches and with a resolution of more than 2000 × 2000 pixels are very expensive," but in the language-game of design-by-doing, we could also express practical understanding of technical constraints and possibilities by "implementing" them in the mockup, prototype, simulation, or experimental situation. Simulations of the user interface were also important in this language-game of design.

As designers, our practical understanding will mainly be expressed in the ability to construct specific language-games of design in such a way that the users can develop their understanding of future use by participating in design processes.

As mentioned above, there is a further important aspect of language-games: We make up the rules as we go along. A skilled designer should be able to assist in such transcendental rule-breaking activities. Perhaps, this is the artistic competence that a good designer needs.

To really learn the language-game of the use activity by fully participating in that language-game is, of course, an even more radical approached for the designer. Less radical but perhaps more practical would be for designers to concentrate design activity on just a few language-games of use, and for us to develop a practical understanding of useful specific language-games of design (Ehn & Kyng, 1987). Finally, there seems to be a new role for the designer as the one who sets the stage for a shared design language-game that makes sense to all participants.

Some Lessons on Design, Skill, and Participation

As in the first practice-oriented part of this paper on designing for democracy at work, I end this second philosophically oriented part on skill-based participatory design with some lessons for work-oriented design.

General lessons on work-oriented design include:

1. Understanding design as a process of creating new language-games that have family resemblance with the language-games of both users and designers gives us an orientation for doing work-oriented design through skill-based participation—a way of doing design that may help us transcend some of the limits of formalization. Setting up these design language-games is a new role for the designer.

2. Traditional "systems descriptions" are not sufficient in a skill-based participatory design approach. Design artifacts should not be seen primarily as means for creating true "pictures of reality," but as means to help users and designers discuss and experience current situations and envision future ones.

3. "Design-by-doing" design approaches such as the use of mockups and other prototyping design artifacts make it possible for ordinary users to use their practical skill when participating in the design process.

Lessons on skill in the design of computer-based systems include:

1. Participatory design is a learning process in which designers and users learn from each other.

2. Besides propositional knowledge, practical understanding is a type of skill that should be taken seriously in a design language-game since the most important rules we follow in skillful performance are embedded in practice and defy formalization.

3. Creativity depends on the open-textured character of rule-following behavior, hence a focus on traditional skill is not a drawback to creative transcendence but a necessary condition. Supporting the dialectics between tradition and transcendence is the heart of design.

Lessons on participation in design of computer-based systems include:

1. Really participatory design requires a shared form of life—a shared social and cultural background and a shared language. Hence, participatory design means not only users participating in design but also designers par-

ticipating in use. The professional designer will try to share practice with the users.

2. To make real user participation possible, a design language-game must be set up in such a way that it has a family resemblance to language-games the users have participated in before. Hence, the creative designer should be concerned with the practice of the users in organizing the design process, and understand that every new design language-game is a unique situated design experience. There is, however paradoxical it may sound, no requirement that the design language-game make the same sense to users and designers. There is only requirement that the designer set the stage for a design language-game in which participation makes sense to all participants.

Beyond the Boredom of Design

Given the Scandinavian societal, historical, and cultural setting, the first part of this chapter focused on the democratic aspect of skill-based participatory design, especially the the important role of local trade unions and their strategies for user participation. In the second part, some ideas inspired by Ludwig Wittgenstein's philosophical investigations were applied to the everyday practice of skill-based participatory design. Practical understanding and family resemblance between language-games were presented as fundamental concepts for work-oriented design.

The concept of language-games is associated with playful activity, but what practical conditions are needed for such pleasurable engagement in design? Is the right to democratic participation enough?

In fact, the experiences from the work-oriented design projects indicates that most users find design work boring, sometimes to the point where they stop participating. This problem is not unique to the Scandinavian work-oriented design tradition. It has, for example, been addressed by Russell Ackoff (1974), who concluded that participation in design can be only successful if it meets three conditions: (1) it makes a difference for the participants, (2) implementation of the results is likely, and (3) it is fun.

The first two points concern the political side of participation in design. Users must have a guarantee that their design efforts are taken seriously. The last point concerns the design process. No matter how much influence participation may give, it has to transcend the boredom of traditional design meetings to really make design meaningful and full of involved action. The design work should be playful. In our own later projects, we have tried to take this challenge seriously and have integrated the use of future workshops, metaphorical design, role playing and organizational games into work-oriented design (Ehn & Sjögren, 1991).

Hence, the last lesson from Scandinavian designs is that formal democratic and participatory procedures for designing computer-based systems for democracy at work are not sufficient. Our design language-games must also be organized in a way that makes it possible for ordinary users not only to utilize their practical skill in the design work, but also to have fun while doing so.

ACKNOWLEDGMENTS

This chapter is reprinted from P. S. Adler and T. A. Winograd (Eds.), *Usability: Turning technologies into tools* (pp. 96–132). New York: Oxford University Press, 1992.

REFERENCES

Ackoff, R. L. (1974). *Redesigning the future*. New York: Wiley.

Aguren, S., & Edgren, J. (1979). *Annorlunda fabriker—Mot en ny produktionsteknisk teori*. Stockholm: The Swedish Employers Confederation.

Berger, P. L., & Luckmann, T. (1966). *The social construction of reality—A treatise in the sociology of knowledge*. New York: Doubleday.

Bjerknes, G., & Bratteteig, T. (1987). Florence in wonderland—systems development with nurses. In G. Bjerknes, P. Ehn, & M. Kyng (Eds.), *Computers and democracy: A Scandinavian challenge*. Aldershot, UK: Avebury.

Bødker, S. (forthcoming). *Through the interface—A human activity approach to user interface design*. Hillsdale, NJ: Lawrence Erlbaum Associates.

Brock, S. (1986). Wittgenstein mellem fænomenologi og analytik. In S. Brock et al. (Eds.), *Sprog, Moral & Livsform*. Århus Denmark: Philosophia.

Brulin, G., et al. (1988). *På rätt spår*. Stockholm: Arbetsmiljöfonden.

Dahlström, E. (1983). *Bestämmande i arbetet—Nøgra idékritiska funderingar kring arbetslivets demokratisering*. Gothenburg: Department of Sociology, University of Gothenburg.

Dreyfus, H. L. & Dreyfus, S. D. (1986). *Mind over machine—The power of human intuition and expertise in the era of the computer*. Glasgow: Basil Blackwell.

Ehn, P. (1989). *Work-oriented design of computer artifacts*. (2nd Edition). Hillsdale, NJ: Lawrence Erlbaum Associates.

Ehn, P., & Kyng, M. (1984). A tool perspective on design of interactive computer support for skilled workers. In M. Sääksjärri (Ed.), *Proceedings of the Seventh Scandinavian Research Seminar on Systemeering*. Helsinki.

Ehn, P., & Kyng, M. (1987). The collective resource approach to systems design. In Bjerknes et al. (Eds.), *Computers and democracy—A Scandinavian Challenge*. Aldershot, UK: Avebury.

Ehn, P., & Kyng, M. (1991). Cardboard computers—mocking-it-up or hands-on the future. In J. Greenbaum & M. Kyng (Eds.), *Design at work*. Hillsdale, NJ: Lawrence Erlbaum Associates.

Ehn, P., & Sandberg, Å. (1979). *Företagsstyrning och löntagarmakt—planering, datorer, organisation och fackligt utredningsarbete*. Stockholm: Prisma.

Ehn, P., & Sjögren, D. (1991). From systems descriptions to scripts for action. In J. Greenbaum & M. Kyng (Eds.), *Design at work*. Hillsdale, NJ: Lawrence Erlbaum Associates.

Ehn, P., et al. (1985). *Datorstödd Ombrytning*. Stockholm: Swedish Center for Working Life.

Einhorn, E., & Logue, J. (Eds.), (1982). *Democracy at the shop floor—An American look at employee influence in Scandinavia today.* Kent, OH: Kent Popular Press.

Floyd, C. (1987). Outline of a paradigm change in software engineering. In G. Bjerknes et al. (Eds.). *Computers and democracy: A Scandinavian challenge.* Aldershot, UK: Avebury.

Foged, J., et al. (1987). *Håndbog om klubarbejde, edb-projekter og nye arbejdsformer.* Århus: HK kommunal.

Fry, J. (Ed.). (1986). *Towards a democratic rationality—Making the case for Swedish labor.* Gower: Aldershot.

Göranzon, B. (1984). *Datautvecklingens Filosofi.* Malmö: Carlsson & Jönsson.

Gustavsen, B. (1985). Workplace reform and democratic dialogue. *Economic and Industrial Democracy, 6*(4), Nov. 461–479.

Hedberg, B. (1980). Using computerized information systems to design better organizations. In N. Bjørn-Andersen (Ed.). *The human side of information processing.* Amsterdam: North-Holland.

Hedberg, B., & Mehlmann, M. (1983). *Datorer i bank.* Stockholm: Swedish Center for Working Life.

Howard, R. (1985). UTOPIA—Where workers craft new technology. *Technology Review, 88.* April, 43–49.

Jackson, M. (1983). *System development.* Englewood Cliffs, NJ: Prentice-Hall.

Janik, A. (1988). *Style, politics and the future of philosophy.* London: Kluwer Academic Publishers.

Kaasbøll, J. (forthcoming). *A theoretical and empirical study of the user of language and computers.* Dissertation. Department of Informatics, University of Oslo.

Kammersgaard, J. (1985). *On models and their role in the use of computers.* Paper Presented at Working Conference On Development and Use of computer Based Systems and Tools. Aarhus University. Denmark. Aug 19–23.

Kosik, K. (1967). *Die Dialektik des Konkreten.* Frankfurt: Suhrkampf.

Kronlund, J. (1978). *Produktionslivets förnyelse—teknik, organisation, manniska, miljö.* Conference in Uppsala. The Swedish Work Environment Fund.

Kubicek, H. (1983). User participation in system design. In Briefs et al. (Eds.), *Systems design for, with, and by the users.* Amsterdam: North-Holland.

Kyng, M., & Mathiassen, L. (1982). Systems development and trade union activities. In N. Bjørn-Anderson (Ed.), *Information society, for richer, for poorer.* Amsterdam: North-Holland.

Lysgaard, S. (1961). *Arbeiderkollektivet.* Stavanger: Universitetsforlaget.

Mumford, E. (1987). Sociotechnical system design—Evolving theory and practice. In G. Bjerknes et al. (Eds.). *Computers and democracy—A Scandinavian challenge.* Aldershot, UK: Avebury.

Nordenstam, T. (1984). Två oförenliga traditioner. In B. Göranzon (Ed.), *Datautvecklingens Filosofi.* Malmö: Carlsson & Jönsson.

Nordenstam, T. (1985). *Technocratic and humanistic conceptions of development.* Stockholm: Swedish Center for Working Life.

Norwegian Employers Federation and Norwegian Federation of Trade Unions. (1975). *General Agreement on Computer-Based Systems.* Stockholm.

Polanyi, M. (1973). *Personal knowledge.* London: Routledge and Kegan Paul.

Qvale, T. (1976). A Norwegian strategy for democratization of industry. *Human Relations, 5.* 468.

Sandberg, Å. (1976). *The limits to democratic planning—Knowledge, power, and methods in the struggle for the future.* Dissertation. Liber, Stockholm.

Sandberg, Å. (1979). (Ed.), *Utredning och förändring i förvaltningen.* Stockholm: Liber förlag.

Sandberg, Å. (1982). *From satisfaction to democratization—On sociology and working life changes in Sweden.* Stockholm: Swedish Center for Working Life.

Sandberg, Å. (forthcoming). *Technological change and co-determination in Sweden—Background and analysis of trade union and managerial strategies.* Philadelphia: Temple University Press.

Seymour, J. (1984). *The forgotten arts—A practical guide to traditional skills*. London: Dorling Kindersley.

Sjögren, D. (Ed.) (1979–83). *Nyhetsblad från Snickeriprojektet*. Stockholm: Swedish Center for Working Life.

Steen, J., & Ullmark, P. (1982). *De anställdas Mejeri*. Stockholm: Royal Institute of Technology.

Swedish Federation of Trade Unions. (1975). *Handlingsprogram för företagsdemokratt och data*. Stockholm.

Swedish Federation of Trade Unions. (1977). *Codetermination on the foundation of solidarity*. Stockholm: Prisma.

Williams, R. (1987). Democratising systems development—technological and organisational constraints an opportunities. In G. Bjerknes et al. (Eds.), *Computers and democracy—A Scandinavian challenge*. Aldershot, UK: Avebury.

Winch, P. (1958). *The idea of a social science and its relation to philosophy*. London: Routledge & Kegan Paul.

Winograd, T., & Flores, F. (1986). *Understanding computers and cognition—A new foundation for design*. Norwood, NJ: Ablex.

Wittgenstein, L. (1923). *Tractatus Logico-Philosophicus*. London: Kegan Paul.

Wittgenstein, L. (1953). *Philosophical investigations*. Oxford: Basil Blackwell & Mott.

Yourdon, E. (1982). *Managing the system life cycle*. New York: Yourdon Press.

5

Achieving Cooperative System Design: Shifting From a Product to a Process Focus

Kaj Grønbæk
Jonathan Grudin*
Susanne Bødker
Liam Bannon
Aarhus University

In the first decades of computer system development, most users of computer systems were engineers and programmers, so "user participation" in development was not actively sought—the developers themselves were good user representatives. In the past 15 years this has changed substantially, as computer use has spread to work environments very unlike the engineering environment (Grudin, 1990). The new divisions of responsibility and the divergence of qualifications have widened the gulf between the developer and user environments. This gulf must be bridged and the most direct approach for doing so is to increase user involvement in development.

Cooperative development—full participation by both developers and users—requires rethinking the tools and techniques used in systems development (Greenbaum & Kyng, 1991). Motives for doing this range from simple cost-benefit arguments—early involvement of users with the future system may lead to adjustment of their expectations, making eventual acceptance more likely—to concern for democracy in working life. Our point here is that user involvement is needed to achieve quality: Better *products,* such as computer applications, will result when the developers have a knowledge about users' practice and future use situations that can only be obtained through cooperation with users (Bødker, 1991; Ehn, 1988; Ehn & Kyng, 1984). This, in turn, entails focusing on the quality of the development process. Although user involvement may take various forms, we believe that in all such situations the users must be taken seriously—involvement cannot be a matter of knowledge extraction or "adjusting their

---

*Presently at the Department of Information & Computer Science, University of California, Irvine.

expectations." It is essential to commit to ensuring benefit for both users and developers.

Good intentions on the part of the developers do not ensure successful cooperation. Structures and processes in a development organization have a large impact on the conditions for user participation. Computer systems development takes place in a wide range of contexts, each with specific advantages and disadvantages for successfully engaging current or potential system users. One commonality is that most development projects are organized around a product to be delivered—they share a *product focus*. In this chapter, we argue for shifting to a *process focus,* wherein a key process element is the enhancement of conditions for user involvement. In the absence of such a shift, individual developers may find themselves engaged in an uphill struggle within their organizations to obtain the user participation needed for project success.

Systems development always occurs within a particular socioeconomic system: Projects are partially framed by the rules and practices of the surrounding society. Many countries have legislated that representatives of different interests be involved in projects; elsewhere, management and labor have agreed to consultation over any proposed change in working conditions. In most European countries formal technology agreements between trade unions and employer organizations establish requirements for development processes. For example, Danish employers initiating a large system development project must first obtain the employees' acceptance of the overall project goals and the expected consequences for work organization. As another example, Danish legislation restricts software products that use databases containing personal information.

Other societal conditions and processes include the availability and qualifications of current and prospective workers and the technology that exists in, or could be acquired by, user and development organizations. General economic conditions are also important; these include customer, supplier, and competitor relationships. Closer to the development project, the structure and processes of development and user organizations can limit or promote the distribution of power and responsibility among different parties (e.g., department managers, individual workers, and unions). As noted by Mathiassen (1981), these conditions are not static, they are changed by organizational and societal processes, possibly including the very development project itself. This is the case, for example, when a development organization hires people with different qualifications or when an application that enables information collection from sensitive databases leads to legislation governing its use.

The interaction of these and other conditions can help or hinder successful cooperation of users and developers. Our discussion focuses on systems *development* projects that are limited in time and resources.[1] Two case studies of devel-

---

[1]Not all system development efforts take place in projects. Exceptions to the project organization include ongoing maintenance work, support of existing systems, and end-user performed programming.

opment projects with different contexts and outcomes illustrate typical obstacles and some opportunities for achieving cooperative design processes. The first case is a product development project in an organizational context that inhibits user involvement. The second case starts as a contract development project based on a product specification, but the project undergoes changes that introduce more active user involvement. This case gives an account of cooperative development in a project marked by a rather complex web of contractual relationships. The projects were chosen both to illustrate different possible contexts for establishing cooperative design processes and because the authors had detailed knowledge about the history of the projects. The projects are discussed in terms of the partners involved and the accompanying agreements or commitments.

Following presentation of the two cases, we discuss several issues that affect successful cooperation: (a) the identification and timing of involvement of the parties to a development project, (b) the nature of the contracts or agreements governing development, (c) how a process focus might be more insightful than the more typical product focus.

As an orienting frame, Fig. 5.1 portrays some of the possible partners in a project and their prototypical relationships. This scheme is utilized in the discussion of our two cases. The outer margins of the figure represent the wider social and political environment in which specific organizational processes, such as system development projects, are embedded. The triangles represent organizations, with overlapping triangles denoting multiple organizations of the same type. The primary relationships between actors are denoted by heavy lines, with bidirectionality implying mutual or reciprocal influence. Dotted lines indicate linkages that are possible but not always present. In fact, the dotted lines represent linkages that we, in this chapter, advocate that development projects establish. The ovals represent the actual groups involved in negotiations between and within each organization.

CASE 1: THE ADVANCED WORKSTATION PROJECT

Note: In this first case there was *no real user involvement in the development process,* other than user testing, which we did not consider as user involvement per se. In our analysis of the project we discuss the apparent rationale for this lack of involvement. Many of the apparent obstacles point to a lack of understanding of the *rationale behind* user participation, as well as to organizational problems in achieving it.

In 1985, a large product development company assembled an elite team of software engineers to develop the company's first powerful, bit-mapped workstation to support "office automation." This project was unusual in having a relatively free design charter and schedule; most projects were revisions of existing products, implementing, under a tight schedule, changes to the product that had been specified in advance. The system was to support word processing, a spreadsheet, business graphics, time management, electronic mail, and other

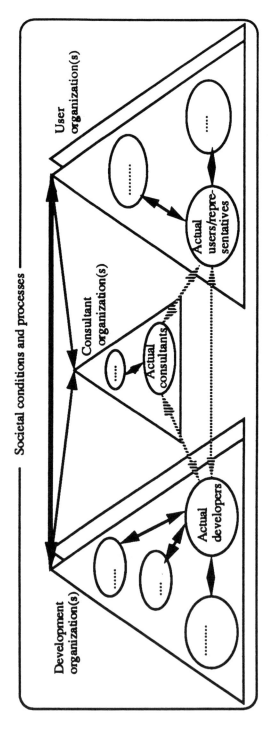

FIG. 5.1. Prototypical partners and relationships in a system development project.

applications to be specified later. Determination of the precise application mix was part of a marketing strategy decided upon at high levels in the company. The human-computer interface was an important aspect of the project. The development team was to define it for both the system-level functions and the applications, the latter to be written by other groups.

The development team had no prior experience in involving users in the design process, apart from informal discussions with developers and secretaries who used the company's products. Most projects provided little time for such activities, few means to accomplish them (the nearest large user organizations were many miles away), and little incentive, in that the changes to be implemented were specified in advance. This project, however, had both a need and an opportunity to break with that practice. Collaboration with prospective users could have benefited the decision between using a proprietary or a commercially available operating system, as well as the design of the graphical interface (new and not well understood at the time). This project had substantial time and resources, but the concept of such collaboration was unfamiliar and was simply never considered. Let us analyze the structures and processes found in this case more closely.

Structure

An outline of the relationships among some of the groups involved in the project is shown in Fig. 5.2. On the left are represented relationships among groups within the product development organization. The arrows from groups in the

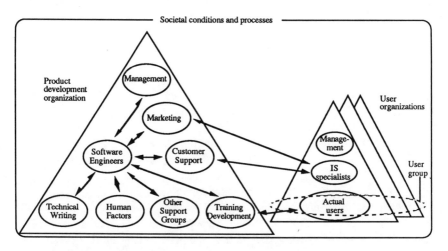

FIG. 5.2.   Project partners and their relationships in the Advanced Workstation project (Case 1).

development organization to groups in user organizations represent commitments related to users of existing products. Similar commitments would be expected to develop for the Advanced Workstation *following its release*. The software engineering group was the hub of activities on this project, not surprisingly. A User Group consisting of representatives of major purchasers of the company's products is represented by the dotted oval on the right.

In recognition of the importance of usability, a User Interface Group of software engineers was established. They were to define the human-computer interface and write much of the code to support it—the window handler, menu handler, message handler, and so forth. An early priority was to write a Users' Manual as a preliminary design specification. A room was reserved and equipped for experiments with prototype use. Recognizing the interlocking aspects of the software, documentation, and training, they obtained two technical writers from the Publications Department, provided space for two performance analysts, and hired someone to design the training and form a bridge to the Training Department. The marketing representative overseeing the project attended meetings from the start. One member of the User Interface Group was assigned to be liaison to the corporate Human Factors group; he also began working to bring in graphic design expertise from industrial design.

The core project development team had only a very general sense of the future users—office workers. Responsibility for thinking about users was sharply divided: Management considered the strategic market segment to pursue; marketing focused on application-level functionality; the development team was responsible for the system's "look and feel"; and support groups were responsible for documentation, training, and other aspects of the system. In fact, the preliminary Users' Manual was actually used to communicate the interface design among these groups and was never reviewed by actual office workers.

Process

This project organization did not survive organizational pressure. The technical writers felt their career paths were jeopardized by their separation from the Publications Department and returned to it, assuming a more distant involvement in the project. The training developer found that senior people in the geographically distant Training Department were unwilling to yield the overall design of training to her and eventually she also left. The performance analysts produced a list of operations and minimum response time requirements for the product based on rules of thumb that the developers found unmotivated and difficult to translate into meaningful design constraints; the performance analysts soon stopped attending meetings. The effort to involve graphic design specialists in designing displays and icons took a full year, due to the need to educate them about the system goals and features and to overcome the resistance of some software engineers, who while inexperienced had grown fond of their own designs. The principal marketing representative focused on the application set; marketing input on other issues was sporadic, based on encounters with design specifications—for example, a previously uninvolved person from International Marketing appeared and demanded that features be changed. The Human Factors group carried out experiments on input device design, windowing techniques, cursor control keys, and other aspects of the system, using internal employees (secretaries or developers) as test subjects for user testing. Their data sometimes helped resolve ongoing arguments, but typically were perceived to cover too little and arrive too late. As developers grew increasingly loathe to "throw away code," a separate prototyping group was established. This worked well until project resources had to be cut back, at which point this group was drafted into development and the prototyping effort stalled.

Contacts and agreements with external groups were minimal. At various times the project received high-level advice from industry consultants and information system specialists in major corporations by showing them the specifications and by bringing them in for round-table discussions (focus groups). At one point, Marketing arranged for several developers to visit clients at a few major customer sites. The developers were astonished at how little the company's products in use resembled those leaving the development company. Customers did considerable modification, adding software or altering the system software so that most users saw only a fraction of the functions. After over a year of development, the company provided a major client with several "prototype systems" consisting of hardware and the most basic system software, entirely for use by the client's in-house development groups. During the course of the project, a major Users' Group meeting was held nearby, but developers were denied permission to attend: Marketing considered it a show for customers.

The project continued for 3 years before being terminated. Development had produced working prototypes of the system (though not of all applications); the concern seemed to be the lack of a market for the product.

Summary and Discussion

As we mentioned at the outset, this project did not really bring prospective users into the development process. Human Factors tested some features on company employees, but we do *not* consider user testing as being cooperative design nor do we consider it to be real user involvement. To some extent, developers regarded themselves as prospective users, since the company had a policy of using its own systems in the development lab. But the system was not targeted for developers or development environments (for example, a spreadsheet was a key application), so those tested were not representative users in many respects—it was this recognition that shocked the developers who saw the modified systems at customers' sites. Nevertheless, marketing representatives, human factors engineers, industrial design engineers, technical writers, training developers, and performance analysts often described themselves as *spokespeople for the users!* In that sense, this project was unusually good at recognizing the need to include these perspectives from the start of the project. Members of these support groups often feel that they become involved on projects too late (Grudin & Poltrock, 1989). In an environment where indirect users' spokespeople often find it difficult to be involved themselves, it is small wonder that no one even contemplates involving actual users.

Would cooperation with potential users have enabled the project to avoid the problems that overtook it? Identifying prospective users is difficult in such a project—which is often given as a reason for not doing so. However, finding a plausible set of uses might have provided a useful, common yardstick for the entire project, and we think that finding such users is possible. A group of users representing some aspects of the prospective use practice is a better choice than none (it is better to have some idea about how the application can be used than none). Resources for this project were substantial and could have supported partial salaries for such users for the time they would need to devote to the project. Furthermore, many user organizations, interested in obtaining state-of-the-art experience with technology, are open to cooperation. Not only is it conceivable that such projects could enlist active user cooperation, doing so might result in a happier fate than met this project.

CASE 2: THE SCHOOL ADMINISTRATION PROJECT

In our second case, both users and developers accepted the need for good relations and worked for the establishment of good conditions for cooperative development. The project concerns a long-term (4 year) effort to develop an administrative system for approximately 100 schools. The description is based on two empirical studies (Holk-Lauridsen & Nielsen, 1989; Klujeff, Møller, & Petersen, 1989) and on direct contacts with developers in the software company involved. The system was planned for an estimated 3000 end-users. Again, we begin by describing the project's structure and processes.

Structure

The system consisted of 10 subsystems to be developed partially in parallel. The system included student administration, building administration, etc. The contract was awarded by the Department of Schools (the customer) to a hardware vendor, who subcontracted the software development. The Department hired a consulting company to manage product testing. Figure 5.3 provides an overview of the partners and their relationships. The heavy horizontal arrow represents the original fixed-price contract between the customer and the hardware company. The arrow linking the management of the hardware and software companies

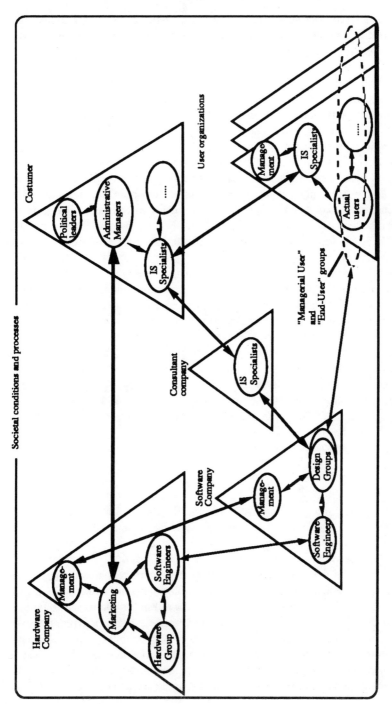

FIG. 5.3. The School Administration Project: Partners and their relationships in (Case 2).

represents a fixed-price subcontract. The arrow linking the software engineers in the hardware and software companies represents an informal communication path that provided mutual support.

The hardware company (together with the software company as subcontractor) had won the competitive bidding with a system sketch defining the subsystem division. The contract provided for a fixed amount to be paid to the hardware company, which was responsible for selecting the hardware and defining the human-computer interface for the software company. A fixed price ($1.24 million) software contract was also established, some of it directly negotiated between the Department of Schools and the software company.

System developer experience with the technology and the application domains varied. Some had trade school experience, others did not. One developer with experience in the application domain was hired for this project. The software developers incorporated their own ideas about the development process into the plans. For example, they insisted that the contract contain a provision specifying the development of prototypes.

The initial terms of the project contained an agreement specifying the involvement of users and their organizations. The agreement contained provisions for user involvement in design and for subsequent user training. A managerial-user group with management level users from six schools was formed and contributed to the overall requirements specification by participating in meetings with the designers. Some managerial-users had prior experience with systems development projects. An end-user group was established, consisting of representative actual users from the schools. They were to test prototypes and be educated to teach the use of the subsystems to the large group of users at each site. The end-user and managerial-user groups received economic compensation for their project work: 400 hours of project work annually per person, which roughly corresponded to the amount of time actually spent.

Process

In the beginning, the system developers and the managerial users worked on specifying the requirements for all the subsystems. This specification process incorporated the use of paper mock-ups of screen images and the development of an initial prototype of the first subsystem. However, when the end-users and the consultants tested the prototype, it was clear that the prototype was incomplete and out of touch with the users' needs. The consultants and the end-users then demanded more influence on the design as a condition for continuing to participate. At this point the project was in serious trouble: The user representatives would refuse to use a system developed in accordance with the specification and initial prototype. At the same time, the software company claimed that major changes in the specification would require renegotiation of the contract. The customer realized that it was partly its fault that the initial specification did not mesh with the actual work carried out in the user organizations. In order not to lose the initial investment, the customer renegotiated the original product contract with the project group (the negotiation is represented in Fig. 5.3 by the diagonal arrow between the project group and the administrative managers). In these negotiations the software company proposed to establish a *process contract,* explained next.

Following these negotiations, the project moved forward under a new project model based on a process contract. A new requirements specification was created for each subsystem by a design group typically consisting of two systems developers, one representative from the consulting company, one managerial user and two or three actual users. The subsystems were developed according to the plan of Fig. 5.4. Each design group worked from the initial overall requirements specification to produce a new specification for the specific subsystem. The managerial users evaluated the specification and identified the changes to the initial requirements specification, which would require modification of the contract. Such changes that were considered to be vital were left in the subsystem requirements specification; those felt to be less important were set aside for possible future development. The cost of these changes was considered in reassessing the price. The software company constructed a prototype and users' guide that covered 80% of the functionality. The adherence of the prototype to the requirements specification and the users' guide underwent a 2-day test by a group consisting of the involved users, some of the managerial users, and the consultants. On the basis of this test, the requirements were again modified, prioritized, and the price renegotiated. The software company then constructed the subsystem, which was tested, corrected, and finally implemented in the user organization. Thus, the plan included two renegotiations of requirements and price for each subsystem. These prototype evaluation experiences contributed to an incremental development process as the implementation of subsystems proceeded. End-users participated in the design groups by producing requirements specifications, evaluating changes to them, and testing prototypes.

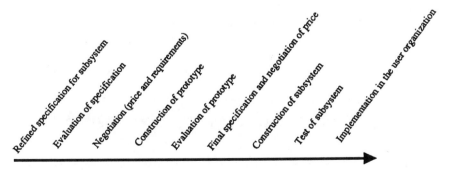

FIG. 5.4.   Plan for each subsystem.

All of the subsystems were developed and the 4-year project came to a close only 11 weeks later than planned. Starting from the original functional specification, 1800 changes were proposed by the user, extending the functionality of the system by approximately 50%. Three hundred and fifty of the changes were postponed. The total price of the system increased by 50% (%0.6 million), but the user got a usable system to support their work.

Summary and Discussion

The initial requirements specification was provided by the customer and refined by systems developers and management-level users. This procedure gave the managerial users considerable opportunity to influence the requirements. However, they did not have a sufficiently detailed knowledge of the actual work carried out by user organizations to specify a usable system. Thus, the prototype of the first subsystem was rejected by the end-users. The rejection led to a renegotiation of the contract, changing it into a contract focusing more on the development process. This led to the creation of design groups with end-user participants applying an iterative prototyping approach. In the design groups, actual end-users assisted the managerial users in revising design documents. This way, users influenced the system design. The managerial users, though, could later remove or postpone certain features for economic reasons. The consultants initially were to participate only by testing prototypes and subsystems, influencing system development only through reports of errors and missing features. However, when the first subsystem failed, they objected to this process. The success of the end-users and consultants in influencing the structure of the project organization was in part because the customer accepted responsibility for problems in the initial faulty, incomplete specification.

The development of later subsystems ran more smoothly. The heavy reliance in the first requirements specification phase on managerial-users, who had a good feeling for the overall system requirements but lacked knowledge of the daily work situation, may explain why testing of the initial prototype resulted in so many suggested changes. For example, paper mock-ups of screen images were made, but not provided to end-users for evaluation. Through subsequent cooperation with actual users, the systems developers overcame this problem and obtained a setting that provided the feedback they needed to develop a good system. In all, 64 (out of 3000) end-users from 32 of the user organizations participated in the design activities and thus contributed actively to the development of the system.

LEARNING FROM OUR CASES: CREATING BETTER CONDITIONS FOR COOPERATION

Our two cases demonstrate various problems in establishing proper conditions for involving users in system development. In the first case, we note the following points: The future users were poorly defined, there was no tradition of direct user involvement in design, and responsibility for thinking about the users was divided. In the second case, the initial emphasis on management-level users was a mistake: They lacked sufficient understanding of the actual work processes. The original product contract constituted a barrier to direct end-user influence on the design. In this section we discuss development conditions and possible ways to overcome the barriers to cooperating with users in such projects. Some development contexts may profit by borrowing techniques that evolved more naturally in another context. We focus the discussion on several issues of particular importance: (a) the identification of users to be involved, (b) the timing of their involvement, (c) the contracts and agreements governing a project, (d) the product focus of most projects.

Late Identification of Partners is an Obstacle to Cooperation

The times at which different potential partners in development are identified— the developers, consultants, and users—is a factor requiring careful consideration in planning for greater user participation in design. The best prospects for a balanced cooperation would logically occur when both developers and users are identified at the outset, as is often true for in-house development projects. For example, a hospital decides that its internal software development group will develop a system connecting nurses and technicians. However, as seen in the outset of Case 2, projects with early identification of both developers and users are not without problems. Choosing representative users is not easy—within a single organization one finds groups with conflicting interests, conflicting perspectives on the organization, and very different possibilities of exercising power over one another (Ehn, 1988; Ehn & Sandberg, 1979). Also, these projects may have few resources to spare. Cooperation can inadvertently antagonize users, especially if they view their own *participation* to be part of a knowledge extraction process that does not help them get influence over their working conditions. In some cases, users are asked to contribute information that is subsequently used to de-skill or even eliminate their jobs. Another danger is of diminishing the likelihood of system acceptance by raising users' expectations unduly. Nevertheless, the early identification of all parties is a favorable condition for cooperation.

The product development environment discussed in Case 1 is quite different. In a "pure" product development situation, the eventual users are not identified until development is complete. The initial product idea does of course include *some* conception of a user population, but specific plans may be closely guarded by management to prevent it from reaching competitors, and the intended or actual use may change along the way. While some market research may precede development, there is little pressure to produce a detailed description of the product's use context. In fact, a vague description may appear to support the desire to appeal to as large a market as possible. This mind-set of striving for broad appeal runs counter to the idea of relying heavily on a handful of users.

For these reasons, in environments in which the development team is clearly identified but user organizations are not, the product focus creates substantial difficulties for full user participation. Even the support groups that see themselves as user advocates, such as technical writing and human factors, are often denied early involvement, due to the perception that development efforts nearer to completion have more urgent need for their attention. Overcoming these obstacles will require changing the development approach to focus on the process, in order to create conditions for user involvement that do not emerge naturally or easily.

Even with a commitment to user participation, identifying the prospective users and the team with which they are to collaborate is not easy in a product development environment. Information about existing and targeted user populations is distributed among people in management, marketing, sales, and customer support organizations. The development team is often a distributed group, with members from software, human factors, technical writing, training development, and elsewhere involved at various times and with varying degrees of communication. On the positive side, these organizations often have considerable motivation to improve the usefulness and usability of their products, they have the resources to commit to it, and they have large numbers of prospective users. (These favorable and unfavorable conditions are covered in more detail in the chapter by Grudin in this volume.)

The Scandinavian UTOPIA project (Bødker, Ehn, Kammersgaard, Kyng, & Sundblad, 1987; Bødker, Ehn, Romberger, & Sjögren, 1985; Ehn & Kyng, 1984) was a research project that applied techniques more typically used in in-house development projects to the design of a system intended for broader distribution. The typographers selected by the graphical unions to cooperate with the researchers were chosen not because they were likely to become actual users of the system, but due to their professional skills and political interest. The typographers continued working part-time at their newspaper to avoid co-option into the technical team and to obtain feedback from a wider group of typographers. Furthermore, the project used newsletters and seminars to reach additional typographers, journalists, newspaper managers, and researchers. Such approaches might be used in commercial settings to explore the possibilities of specific application areas, narrow or wide, and to provide feedback much earlier than is available through alpha or beta site testing, although sensitivity about disclosing plans may be an inhibiting factor.

A different problem with the timing of involvement arises when development follows competitive bidding. Here, too, we find a natural product focus: the requirement to specify the system to be built. In this case, the customer/user organization is first involved. It may have to define the system without input from the eventual developers, although perhaps with the help of consultants, as in our second case study. The developers may not be able to modify or "break" the contract as part of an interactive design process and may even be discouraged or prohibited from establishing ties to the user organization for reasons of security or for fear of incurring charges of favoritism on subsequent contract bidding. In our second case the initial contract terms were modified substantially due to problems that occurred with the initial product specification, but contractual arrangements are often more inflexible.

For large competitively bid contracts, the motivation and resources to overcome these obstacles may exist. The next section discusses a key element in overcoming the obstacles—the development of more flexible, process-oriented contracts. (The influence of timing of involvement in development on user participation is discussed in more detail in Grudin, 1991.)

Fixed Contracts are Obstacles to Iterative Design

In Case 2, competitive bidding led to a fixed product contract between the development organizations and the customer/user organizations. The contract assumed that the development of the system could be based on an initial specification made by a group of management level users together with the developers, but, as described on p. 88, the customer recognized the need for renegotiating this in order not to lose the initial investment. Thus, it was possible to set up a new project model that included price renegotiation, iterative design, and incremental delivery of subsystems. Because some system feature ideas that emerged from direct user involvement would be expensive to implement, it was agreed to renegotiate the price based on their assessment of the prototypes. This case suggests that an initially disastrous design proposal can evolve into a successful development project when a fixed contract is replaced with a strategy providing for flexible renegotiation.

The issues raised by fixed contract proposals are too complex to take up fully in this chapter. It is an important issue, because changing socioeconomic conditions may bring about an increased reliance on this approach to software acquisition. In discussing problems for cooperative system development in contract situations, Gundry (1988) claims that "co-operative system acquisition" is losing ground to competitive purchasing due to the difficulties of controlling money and time with the former policy. He is pessimistic about the possibility for extending the flexibility of contracts and argues for finding a way to fit user concerns into the more restrictive competitive type contracts, although he recognizes the difficulties and limitations in doing so. We do not share his pessimism about more cooperative strategies and more flexible process contracts, having seen encouraging signs within the commercial sector that such contracts can be negotiated successfully.

In general, fixed contracts may hinder iterative design, independent of the type of project considered. Development contracts should be shaped as process contracts between customer and development organizations with scheduled negotiation activities. The contract should outline a set of work tasks which the customer wishes to have improved, together with a set of guidelines or procedures about how, rather than attempt to define the system in advance. Of course, customers prefer the security of having a fixed price/fixed time contract, but the reality is that such contracts often live in fiction. The deliverable is generally a far from adequate simulacrum of the desired item. The potential magnitude of this problem is seen in a United States General Accounting Office study of nine software development projects (which were not selected at random, however), where under 5% of the overall expenditure resulted in software that could be used as delivered or with minor changes (Martin, 1988, p.4). Agreeing on a process to take place within a certain frame can easily create equal or better security.

Where possible, contracts might incorporate aspects of the project model that evolved in our second case. Boehm (1988) proposes that in-house development projects incorporate iterative design through cyclic or spiral models that explicitly schedule checkpoints for reconsidering future activity. For contract development projects, he describes experimental approaches for increasing flexibility, including multistage contracts in which several developers produce prototypes in the initial round, level-of-effort and award-fee contracts for evolutionary development, and design-to-cost contracts. Project models such as the spiral model seldom define the degree of user involvement, hiding it under the label of phases such as "prototype evaluation" or "test." However, we suggest that more active user involvement, at least to the extent seen in Case 2, is more likely to produce usable interactive systems. Thus, contracts between development and user organizations could also provide explicitly for the use of certain cooperative design techniques (as described in Bødker & Grønbæk, 1989; Greenbaum & Kyng, 1991; Kyng, 1989) to facilitate active user involvement at different stages.

A Product Focus is too Narrow

Cases 1 and 2 deal with development tasks where the user and development organizations are separate units with no history of cooperation. The projects start out trying to specify the product to be developed. In this sense they are very similar to traditional system development projects that start by attempting a *complete* product specification, with the inherent assumption that the future users' needs can be completely analyzed and anticipated. This is a strategy, however, that leads to many products being rejected by users (Bødker, 1991; Ehn, 1988).

Our Cases 1 and 2 diverged from the traditional approach to varying degrees in recognizing a greater need for focusing at the outset on the needs of potential or actual users. Case 1, the workstation development project, provided only a modest change of focus, primarily through greater communication among those responsible for different aspects of the user interface. A more radical shift of focus was seen in Case 2, the school administration project. There, the development project changed from a fixed contract for the delivery of prespecified software to an evolutionary design process with continual, active user involvement. Time was invested in establishing coordination patterns that would not have formed without explicit attention to the processes and conditions required for successful cooperation. This investment in the process was the main reason for the successful outcome.

There are encouraging signs of a shift away from a simple product focus toward an understanding of the process of system design, development, use, redesign, and reuse. Breaking down (or tightly coupling) the development/use distinction requires rethinking the systems development process and the nature of the commitments involved. A number of voices, from different backgrounds, can be heard in support of this shift.

For example, the designer Chris Jones (1988) criticizes software designers for focusing too much on the use of the completed product. He argues for more attention to the process as an end in itself, not simply a means. Regarding the standard procedure of building to a fixed set of specifications, Jones notes: "Both parties (designers and clients) have to give up the use of the requirements as a semi-legal basis for control and measurement and agree to work together in the continuous meta-process of evolving the design brief and sharing in the eventual decision as to how the problem is to be seen and solved." Continuing, he notes: "Functions, statements of requirements, are essential but temporary. Without them we cannot begin, but unless we can change them we cannot finish, cannot discover. ." Jones concludes with the strong claim that ". . . [we must] recognize that the 'right' requirements are in principle unknowable by users, customers, or designers at the start." This position calls into question the nature of most formal contracts today. Similarly, the consultant Tom Gilb (1990) stresses the need to focus on process, not method or static product. He notes that current development methodologies ". . . are based on a static product model. They do not adequately consider our work to be a continuous process—derived from the past and being maintained into the future."

Yet another voice in support of this shift, coming from academic software engineering, is that of Floyd (1987). She argues for more emphasis on the process of software development than on the efficiency of the resulting code: "The product-oriented perspective regards software as a product standing on its own, consisting of a set of programs and related defining texts . . . considers the usage context of the product to be fixed and well understood, thus allowing software requirements to be determined in advance," while the process-oriented perspective "views software in connection with human learning, work and communication, taking place in an evolving world with changing needs . . . the actual product is perceived as emerging from the totality of interleaved processes of analysis, design, implementation, evaluation and feedback, carried out by different groups of people involved in system development in various roles." This shift in perspective calls into question the separability of development and use and the concept of a "complete" requirements specification. Again, Gilb (1990) admonishes software developers to accept that requirements are always changing and to focus on handling this situation by catching problems early through user involvement in the design and specification phase. He also stresses the need for prototyping and frequent iteration. Taking this view for granted, the negative consequences of obstacles to cooperative design, as exist in many existing development contexts, is apparent. In contexts lacking patterns of cooperation between users and developers, a deliberate effort is required to change the conditions, but there is also a good chance that the effort will pay off.

CONCLUDING REMARKS

We have described some obstacles to cooperative design, i.e., design with active user involvement, which can be found in many types of system development context today. Effort and expense are required to find and implement techniques to overcome them—to develop methods and tools, and to change organizational practices and perspectives. We have argued that the concern for quality products and processes is the driving force: To create better computer applications, developers need to be more concerned with the work processes in which the computer systems will eventually be used and with the outcome of these processes. Traditional ways to obtain knowledge about work processes, such as user modeling and system description, have failed (see Bannon & Bødker, 1989; Bødker, 1991). We argue for greater emphasis on user–developer cooperation in the development process. Such cooperation requires early identification of the project partners and flexible contracts. Contracts should focus on creating a development process based on user–developer cooperation rather than on attempts to specify products fully in advance. In addition, many of the costs of providing more usable systems are decreasing—computer processing time, memory, maintenance, and so forth. Thus, both market pressures and concern for the use of technology on behalf of those working with it are leading in the same direction. We have discussed apparent obstacles to these positive changes and some attempts to overcome them. Although these attempts were not uniformly successful, we believe that the approaches, described in this and other chapters of the volume and in Greenbaum and Kyng (1991), show promise for many system development projects. They are the first steps towards a radical reevaluation of current system development procedures and processes. This reevaluation will lead to increased emphasis on cooperative design as well as greater emphasis on the process per se.

REFERENCES

Bannon, L., & Bødker, S. (1991). Beyond the interface: Encountering artifacts in use. In J. Carroll (Ed.), *Designing interaction: Psychological theory at the human-computer interface* (pp. 227–253). New York: Cambridge University Press.

Boehm, B. (1988). A spiral model of software development and enhancement. *IEEE Computer, 21*(5), 61–72.

Bødker, S. (1991). *Through the interface–A human activity approach to user interface design.* Hillsdale, NJ: Lawrence Erlbaum Associates.

Bødker, S., & Grønbæk, K. (1989). Cooperative prototyping studies—Users and designers envision a dentist case record system. In J. Bowers & S. Benford (Eds.), *Studies in computer-supported cooperative work: Theory, practice and design* (pp. 315–332). Amsterdam: North-Holland.

Bødker, S., Ehn, P., Kammersgaard, J., Kyng, M., & Sundblad, Y. (1987). A Utopian experience. In G. Bjerknes, P. Ehn, & M. Kyng (Eds.), *Computers and democracy—A Scandinavian challenge* (pp. 251–278). Aldershot, UK: Avebury.

Bødker, S., Ehn, P., Romberger, S., & Sjögren, D. (Eds.) (1985). *Graffiti 7. The UTOPIA project: An alternative in text and images.* Stockholm: Arbetslivcentrum.

Ehn, P. (1988). *Work-oriented design of computer artifacts.* Falköping: Arbetslivcentrum/ Almqvist & Wiksell International.

Ehn, P., & Kyng, M. (1984). A tool perspective on design of interactive computer support for skilled workers. In M. Sääksjärvi (Ed.), *Proceedings from the Seventh Scandinavian Research Seminar on Systemeering* (pp. 211–242). Helsinki: Helsinki Business School.

Ehn, P., & Sandberg, Å. (1979). *Företagsstyrning och löntagarmakt [Management control and wage earner control].* Prisma, Stockholm.

Floyd, C. (1987). Outline of a paradigm change in software engineering. In G. Bjerknes, P. Ehn, & M. Kyng (Eds.), *Computers and democracy—A Scandinavian challenge* (pp. 191–212). Aldershot, UK: Avebury.

Gilb, T. (1990). Project management for the 1990s. *The American Programmer,* 16–30.

Greenbaum, J., & Kyng, M. (Eds.). (1991). *Design at work: Cooperative design of computer systems* Hillsdale, NJ: Lawrence Erlbaum Associates.

Grudin, J. (1990). The computer reaches out: The historical continuity of interface design. *Proceedings of the CHI'90 Conference on Human Factors in Computing Systems,* 261–268. New York: Association for Computing Machinery.

Grudin, J. (1991). Interactive systems: Bridging the gaps between developers and users. *IEEE Computer, 24*(4), 59–69.

Grudin, J., & Poltrock, S. (1989). User interface design in large corporations: Communication and coordination across disciplines. *Proceedings of the CHI'89 Conference on Human Factors in Computing Systems,* 197–203. New York: ACM.

Gundry, A. J. (1988). Humans, computers, and contracts. In D. M. Jones & R. Winder (Eds.), *People and computers IV* (pp. 161–175). Cambridge, UK: Cambridge University Press.

Holk-Lauridsen, M., & Nielsen, H. B. (1989). *Designprocessen som et sprogspil—belyst gennem et empirisk studie af 3 designprojekter* [The design process as a language game; seen through an empirical investigation of three design projects]. Unpublished Masters Thesis, Aarhus University.

Jones, J. C. (1988). Softecnica. In J. Thackara (Ed.), *Design after modernism: Beyond the object* (pp. 216–226). London: Thames & Hudson.

Klujeff, R., Møller, E. M., & Petersen, K. V. (1989). *The role of users in systems development.* Unpublished Master's Thesis, Aarhus University.

Kyng, M. (1989). Designing for a dollar a day. *Office: Technology and people, 4*(2), 157–170.

Martin, C. F. (1988). *User-centered requirements analysis.* Englewood Cliffs, NJ: Prentice-Hall.

Mathiassen, L. (1981). Systemudvikling og systemudviklingsmetode [Systems development and systems development method]. *DAIMI PB, 136.* Århus: University of Aarhus.

6 Obstacles to Participatory Design in Large Product Development Organizations

Jonathan Grudin
University of California, Irvine

The need for the developers of interactive systems to understand the eventual users and the work that the systems are to support is well known. How do developers obtain this understanding? They rely on intuition. They learn about users indirectly, by reading or being told about users' needs. Sometimes they study or interact with users in constrained situations. Participatory design is a response to the recognition that these methods are often unsatisfactory—the utility and the usability of software could be markedly improved. Although good interactive systems exist, the difficulty of developing them is substantial. Many existing successes are the result of an undesirably long evolutionary process.

This chapter explores the underlying problems in one systems development context: large product development organizations. Most of these companies were formed before product usability attained its present visibility. As a result, they gave no consideration to the particular needs of interactive systems development when defining their basic organizational structures and development processes. Several specific difficulties in achieving and benefiting from user participation in development are frequently found in these organizations. Eventually, organizational change may be required to overcome the constraints and forces confronting developers. In the meantime, those working within such organizations must be aware of the problems and seek constructive paths around them.

One goal of participatory design is to understand organizational change in computer *use*—the effects of introducing technology on organizational structure and process and the effects of organizational restructuring on the way work is carried out. Readers of this chapter are invited to examine the same issues in the computer *development* context: by reflecting on the effects of contemporary organizational structures and processes on product development, and by reflecting on how organizational change might eliminate the obstacles created by existing structures and processes. Widespread introduction of participatory design would have major organizational effects on a product development company.

This chapter draws on the growing literatures in two fields: human-computer interaction, much of which originates in product development companies, and participatory or collaborative design, most of which is drawn from other development contexts. It also relies on surveys and interviews of over 200 interface designers in several product development companies (Grudin & Poltrock, 1989; Poltrock & Grudin, in press), experiences in product development, and thousands of conversations with fellow developers over the years. Of course, organizations vary considerably. Reliable, industry-wide data are difficult to find.

The obstacles described here are encountered, but not universally. The hope is that the forewarned reader will be better able to anticipate, recognize, and respond to these and similar challenges if and when they arise.

THE CONTEXT: PRODUCT DEVELOPMENT PROJECTS AND MARKETS

This section outlines the development context of the obstacles to user participation discussed in the rest of the chapter. First, large product development organizations are identified and described. Then, development projects within these organizations are analyzed, revealing why the human-computer interface or "user interface" is the principal focus of utility and usability for a product development team. Finally, the range of possible user involvement, from full collaboration to occasional or indirect consultation, is discussed.

Product Development Organizations

This chapter focuses on large organizations that develop and market "off the shelf" or "shrinkwrap" software applications and systems. This excludes in-house development projects and projects undertaken to fulfill contracts. Internal and contract development have different advantages and disadvantages for user involvement in developing interactive systems (Grudin, 1991a). Of course, companies often straddle categories: a product development company that bids on government contracts, a company that markets a system built initially for internal use or under contract, and so forth. In addition, *small* product development companies may not experience the problems described here, and companies of moderate size are likely to experience some and not others.

Product development accounts for only a fraction of interactive systems development, but it is an influential fraction. Large product development companies are visibly concerned with usability and "look and feel." They hire and train many user interface specialists, recruiting heavily from research universities.

These specialists dominate the conferences and journals in the field of human-computer interaction, especially in the United States, as reflected in the strong presence of product developers from IBM, Digital, Hewlett-Packard, Xerox, Apple and other large companies at Computer and Human Interaction (CHI) Conferences. At the 1990 Participatory Design Conference in Seattle, approximately 25% of those registering were employees of large product development companies.[1]

Most of these companies matured in the 1960s and 1970s. Their principal source of revenue was selling or leasing hardware; software functionality was secondary and the human-computer interface received little attention. Most processing was batch, not interactive, and the immediate users were computer professionals. Since then, software has come to rival hardware in importance—many of the successful new product development companies of the 1980s primarily sell software. Until the success of the Macintosh in the late 1980s, the focus was entirely on functionality and price; now, the interface is increasingly important.

Although attitudes are changing, the business operations and development practices of most of today's large product development companies were formed when hardware and software functionality were the only considerations. It is not surprising that their basic organizational structures and processes do not facilitate interface development. In fact, we will see that the design and development of good interfaces are often systematically *obstructed*—not intentionally, but obstructed nonetheless.

Development Projects: Defining Function Before Form

The process of creating a product can be separated into two parts: the events before and after the development team is formed. The timeline of an ongoing development project includes a start date, a projected completion date and a number of milestones. Reality can be fuzzier—one event flows into another and decisions are gradually recognized to have emerged along the way. Still, the project start date does have meaning—a team is formed, assignments announced, and budgets allocated—and may even be dramatized, as when the first working name of a project is a number based on the month and day of its initiation.

Product definition, the first phase, *precedes* the project start date and the development team formation. During this phase the high-level "functionality" is identified. In the second phase, the team is assembled and it designs and develops the necessary low-level operations, the visual appearance, the documenta-

---

[1]Approximately 35% were from academic institutions, 20% from companies in other businesses likely to have internal systems development, and the remaining 20% included consulting organizations, small software development companies, government, media, etc.

tion, and other aspects of the human-computer interaction. Loosely speaking, the functionality is defined in the first phase and the interface in the second.

It is notoriously difficult to draw a line between software functionality and its "user interface." Most participatory design projects have very sensibly avoided this issue by considering the system as a whole, to the extent possible. After all, while one could have a usable system that is not very useful or a useful system that is not very usable, the most desirable outcome is clear, so it makes sense to work on both together.

Computer users are generally more interested in the work that the computer is supporting than in the interface, as captured in the title of Bødker's (1991) book *Through the Interface.* Recognizing the importance of system functionality, the Europeans involved in these projects (as well as many Americans who also focus on the in-house development or the internal use of information systems) are leery of the "user interface focus" of conferences such as the ACM SIGCHI Conferences.

This is a key difference between the product development environment and the in-house development context typical of participatory design experiments. In the latter, management may have preconceptions about the function of the system to be developed; resisting a premature definition is necessary in order to consider the entire work and technology situation. But in product development, the situation is far more extreme. The high-level functionality is defined before the development team is formed, by a different group of people, based on a wide range of factors. The development team is handed the product idea and is responsible for the remaining design. This division of responsibility, experienced by every project, maps loosely into the functionality vs. interface distinction and explains the restricted range of issues addressed at CHI Conferences, which are dominated by product developers.

The precise division of responsibility varies from project to project. In fact, as the interface grows in importance, key design decisions move from the development phase to the product definition phase. A product is specified to have a graphical interface, adhere to a corporate "look and feel," run on Windows 3.0, and so forth—constraining the design space of the development team.

Where can participatory design enter this two-phase process? First, consider the possibility of user participation in product definition. This phase is generally carried out by a management group. Companies vary in the degree to which they are driven by engineering or marketing initiatives, but marketing typically influences product definition. "User needs" or marketplace demand is an important factor. Open-ended needs-finding—involving users without any preconceptions—is not unheard of, but it is rare; ideas for products or enhancements are usually plentiful. Other factors come into play. Does the potential product fit coherently into the existing product line? Will it undermine sales of other products? Are major competitors using a feature to make inroads into a company's existing customer base? Do the marketing and sales force have the expertise and motivation to sell it? Will sales be enough to pay for the necessary advertising campaign? Does it test a potentially desirable new application domain for the company? Is it a new release needed primarily to reassure customers that the product line is not being abandoned? Strategic decisions such as these are crucial to the company and are often closely guarded for competitive reasons. Although major customers are consulted, the possible role of user involvement at this point is very limited. As the product definition phase proceeds, issues are addressed that might benefit from such involvement. This can lead to market surveys and focus groups. Often mediators are relied upon for information about potential users—consultants, the trade press, internal marketing staff, users' "representatives" (e.g., systems analysts, who also represent the possibly conflicting interests of others, such as the users' management). But "end users" from customer organizations are unlikely to be full participants in this complex product definition process. In fact, rarely is an individual *developer* from the *same* organization fully involved.

The rest of this chapter focuses on the period after the baton is passed to the development team. Once the project team is assembled, the product definition group largely recedes from view, moving on to other concerns while monitoring progress through documentation and management reviews. This strongly phased process is the reality to be addressed. Viewed as a candidate for possible organizational change, this approach to development resembles the requirements-driven model that originated in contract development and is widely applied to internal development (Grudin, 1991a). The belief that a written requirements specification is enough to communicate a product idea had more validity for noninteractive systems and may yield the problems found in contract and internal development: A product is delivered that is not quite what those defining it had in mind or what its recipients find usable. As product usability draws more attention, more rapid-prototyping efforts occur within marketing groups prior to product definition, permitting more opportunity for collaboration among marketers and developers—and perhaps users.

The part of the design of interest to the eventual users that remains to be defined by the development team corresponds approximately to the human-computer interface, liberally defined. This includes documentation, training, hot-line support, and other elements that directly affect the users' experiences with the product. It also includes low-level functions; for example, whether a one-step "move" or a two-step "cut" and "paste" is provided. The development team may have license to add features that can be justified. Conversely, as noted above, some aspects of the interface may have been constrained in the product definition phase.

To summarize, while there is generally room for innovation and new product definition, the nature of product development works to separate the definition of high-level functionality from subsequent development. For this reason, user participation in product development is overwhelmingly in furtherance of the design and evaluation of the interface.

Degrees of User Involvement

What is the optimal degree of user participation in development? If you are developing a compiler, the need for users' involvement is minimal. If you are copying features from an existing product in a mature application area, limited contact with potential users can be adequate. If you are developing an interactive system in a new domain, full collaboration with users can be essential.

Again, we find an important contrast to the in-house development situation found in many participatory design efforts. In-house projects involve the development of a new system for a specific group of users. Since the system must be accepted by the target group, it makes sense to learn as much as possible about the future users' shared or even individual backgrounds, work practices, and preferences. The product situation is different—the specific users are not known in advance, one is in a sense targeting a "greatest common denominator." Earlier versions or competitors' products often serve as guides, and as noted earlier, many of the major decisions were made in the product definition phase.

Product developers also have a particularly wide range of *indirect* approaches to acquiring *some* understanding of computer users. They learn about users in courses, conferences, and trade shows. They rely on marketing and sales people, customer service and training staff, consultants, information systems specialists, users groups, standards organizations, trade magazines, and journals. The adequacy of these and other indirect approaches depends on the circumstances, but many of these mediators exist to serve this communication role and have an incentive to make themselves useful.

Although there is no universal answer or easy algorithm for calculating the optimal degree of direct user involvement, current practice seems to produce too little. Many products are unusable or unnecessarily difficult, and strong forces are pushing developers to obtain more detailed information about future users of their products. Four of these forces are:

1. The spread of computer use into environments that are increasingly unlike development environments. In the past, most computer users were engineers or programmers, so developers' intuitions about user environments were good. As computer users become less technical and more diverse, developers acquire a corresponding need to obtain more information about them and their work environments.

2. The rising expectations of computer buyers and users. Even as the growing capabilities of applications and systems present greater challenges in instruction and information presentation, the willingness to adjust to the system and the

tolerance for poor interfaces are declining. As prices decline, heavy investment in training is less palatable. Again, the burden shifts to the developers to create responsive systems.

3. Maturing application areas. In a new market, functionality is likely to govern purchasing decisions. But when several products offer comparable functionality, ever more detailed knowledge about users is needed to fine-tune products and provide an edge.

4. Emerging applications that support groups ("groupware") require more knowledge of users and their environments than did single-user products. Groupware must support a wider range and greater percentage of users in a given setting, bringing product developers closer to the situation that in-house developers have faced. The limitations of any one person's intuitions for group behavior and a greater need to consider issues surrounding adoption in the work environment require developers to gather more information than before (Grudin, in press).

These escalating demands, on top of frequent product failure and user dissatisfaction, suggest that intuition and indirect methods of learning about users, which are not doing well now, are likely to be increasingly inadequate. So we turn to consider direct user involvement.

Direct contact takes many forms. Actual or potential users are observed, surveyed, interviewed, tested, studied, and experimented upon. Remarkable ingenuity has been displayed in finding ways to poke and prod users. Are approaches short of full collaboration enough? In in-house development projects, charged with addressing the specific needs of a well-defined set of users, probably not. In product development, they can be enough to answer *some* questions. Product developers, aiming for a broad market, are less concerned with differences than with commonality. They have focused on characteristics shared by most users: motor skills (keyboard layout, mouse control), perceptual processes (character legibility, color contrasts), and cognitive processes (recognizing menu item names, scanning displays for information). Limited forms of user involvement such as formal experiments resolve some of these questions. However, to deal with issues closer to the task or application domain, these laboratory methods are less adequate. In fact, even the low-level design issues listed earlier can turn out to hinge on aspects of the work situation (Grudin, 1989).

In many situations, a full collaboration with future users seems vastly preferable to attempts to "extract information" from users late in development. Consider the situation turned around: Imagine a setting in which the *developers* play a secondary role. There is a very "tailorable" application and the "project team" consists primarily of users, with programmers in a support role. In this example, we might argue that the team of users should attempt to keep the programmers involved from the start and not hope to rely on "extracting information" from them when some coding is necessary in the late stages of redesigning the system!

Product development companies are not entirely unaware of the virtues of participatory design; for example, "domain experts" from user organizations are hired to work on development teams. Gould and Lewis (1983) of IBM made an early, forceful argument for participatory design, eschewing reliance on mediators and more limited empirical approaches: "We recommend that *typical users* (e.g., bank tellers) be used, as opposed to a 'group of expert' supervisors, industrial engineers, programmers. We recommend that these potential users become part of the design team *from the very outset* when their perspectives can have the most influence, rather than using them post hoc to 'review,' 'sign off on,' 'agree' to the design before it is coded." This advice has been prominently republished (e.g., Gould, 1988; Gould, Boies, Levy, Richards, & Schoonard, 1987; Gould & Lewis, 1985) and is widely cited, yet it is rarely followed. Gould and Lewis (1985) allude to unexplored "obstacles and traditions" that stand in the way. This chapter explores those obstacles and traditions.

OBSTACLES TO USER INVOLVEMENT

This section covers challenges to user involvement that arise in several ways. Some are due to the inherent nature of product development: The actual users are not truly identified until development is complete and the product is marketed; potential users work for different organizations; any one set of users may be too limiting. Additional obstacles can be traced to the division of labor within the development organization. Typical allocations of responsibility serve useful purposes, but they distribute aspects of the interface across organizational boundaries and separate software developers from the world outside (see Fig. 6.1).[2] With noninteractive systems this was not a problem, but it is a problem now that is likely to get worse. Contact with customers and users is the province of groups or divisions outside of development: sales, marketing, training, field support, and upper management. The people assigned these tasks are not primarily concerned with the interface, their relevant knowledge is not systematically organized, and they are often located far from the developers. They have a limited sense of what information would be useful or to whom to forward it. Finally, after discussing these structural impediments, we examine difficulties that can be traced to standard software development procedures and techniques.

Challenges in Motivating Developers to Work With Computer Users

Successful user participation requires most or all members of the development team to commit to user involvement. One person can work with users and try to introduce the results into the process, but iterative development requires a broad

---

[2]Many small and large variants of this organizational structure are found. To take one example, functions such as Quality Control and Performance Analysis may be handled centrally.

A Typical Organizational Structure

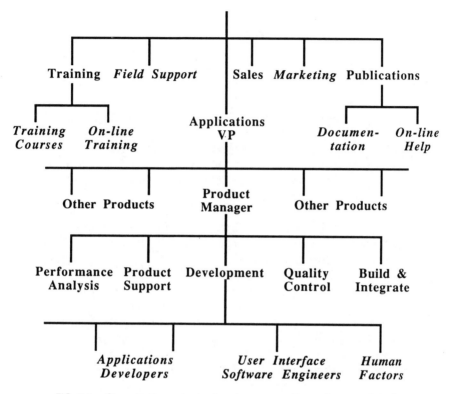

FIG. 6.1.   Organization chart showing separation of user-related functions.

commitment; prototyping and testing generally require software support; management must be willing to invest the resources; the help of others may be needed to initiate contacts with users; and the results have to be valued.

Although most developers would agree to user involvement in principle, it requires a greater commitment to make it work. A lack of follow-through may occur for several reasons. Engineers may lack empathy or sympathy for inexperienced or nontechnical computer users. Developers and users may discover that they have different values, work styles, and even languages, hindering communication. Developers tend to be young, rationalistic, idealistic, and the products of relatively homogeneous academic environments. They often have little experience or understanding of the very different work situations and attitudes of many system users. The best of intentions can succumb to these factors, especially in the face of the slowness and imprecision that often accompany user involvement.

Challenges in Identifying Appropriate Users

Developers may have a market in mind, but the actual users of a product are not known until the product is brought. The fates of many products, both negative and positive, are reminders of the inherent uncertainty in product development. The IBM PC is an example of a product that had a wider appeal than expected, while we can be confident that the designers of countless failed products anticipated users who never materialized.

Further obstacles to identifying potential product users stem from the desire to appeal to a broad range of people. The effort is focused on casting as wide a net as possible; reversing gears to try to identify specific or characteristic users is difficult. Choosing one seems to eliminate others. The seriousness of the problem of defining characteristic users is reflected in the experience of Scandinavian participatory design. Even with the advantage of beginning with relatively constrained user populations, within one industry or even one organization, selecting representative users could be a major challenge (e.g., Ehn, 1989, pp. 327–358). Such problems are greater for developers of generic products.

The division of labor also thwarts developer-user contact. User interface specialists rarely have "the big picture." They often work with a development team assigned to a single application or even to part of an application. Not even the project manager has a perspective encompassing the application mix that customers will use, the practices and preferences of the installed customer base, and strategic information about the intended market for a product. A broader perspective is found in Marketing or Sales divisions—which can be geographically and organizationally distant from the development groups. The projected market—the identity of the future users—may be closely guarded by upper management due to its competitive importance.

In large companies, marketing and sales representatives become species of users of products emerging from development. They also consider themselves to be internal advocates for the customers. Since the customers are often information specialists or managers, rather than "end users," the chain of intermediaries lengthens. Low levels of contact and mutual respect between marketing and development can further reduce the value of this very indirect link between developers and users (Grudin & Poltrock, 1989; Mitch Kapor face to face, 1987; Poltrock & Grudin, in preparation).

Another complication in identifying appropriate users is that a system is often modified substantially after the development company ships it but before the users see it. This is done by software groups within customer organizations and by value-added resellers who tailor products for specific markets, for example. These developers are in a real sense users of the product—perhaps among the most important potential users. It may be more appropriate for *them* to involve the actual end users. In any case, the initial development team must discover which aspects of their design are likely to be passed through to users. Third-party intermediaries represent an opportunity, but their role also complicates the selection of representative end users.

Challenges in Obtaining Access to Users

Once candidates have been identified, the next challenge is to make contact. Obstacles may arise within either the users' organization or the development organization.

Contacts with customers are often with managers or information system specialists, rather than with the computer users themselves. Getting past them may not be easy: Their job is precisely to represent the end users. In addition, the employers of prospective users may see no benefit in giving them time off to work with an outside development group.

Within the product development company, *protecting (or isolating) developers from customers is traditionally a high priority.* The company cannot afford to let well-intentioned developers spend their time customizing products for individual users—their mission is to develop generic improvements to benefit scores or hundreds of users. Savvy customers are well aware of the value of having the phone number of a genial developer. And barriers erected to keep users from contacting developers also prevent developers from easily connecting with users: The relationships and channels are not there.

The development company's sales representatives may be reluctant to let developers meet with customers. A developer, coming from a different culture, might offend or alarm the customer, or create dissatisfaction with currently available products by describing developments in progress. Similarly, Marketing usually considers its role to include informing the development organization about customer needs and may fear the result of random contacts between developers and users. In one company, developers, including Human Factors Engineers, were prevented from attending the annual Users' Group meeting. Marketing viewed it as a show staged strictly for the customers.

Sometimes, perhaps too often, potential users are found *within* the development company. This is convenient, but a dangerous special case to rely on. The company is not in business to build products for itself, and user environments resemble development environments less each year.

Challenges in Motivating Potential Users

In an in-house development project, the product developers share the same management (at some level) with the potential users. This is not true in the case of product developers and external users, making it more difficult for users to obtain time away from their jobs. In addition, the potential users can be less motivated, knowing that they may not become actual users of the final product. The problems of sustaining user interest are substantial even in internal development projects (e.g., Bødker, Grønbæk, & Kyng, in this volume). Of course, for contacts of limited duration, many computer users and their employers are pleased to be consulted. But the successful internal development project described in Grønbæk, Grudin, Bødker, and Bannon (in this volume) obtained 400 hours per participant user per year, an investment few customers will make in product development.

Potential users can be less motivated if they do not see how the planned product would benefit them personally. This is particularly a problem for groupware applications, most of which require widespread use but selectively benefit managers (Grudin, 1990). In the worst case, the potential users feel that their jobs are threatened by a product that increases overall efficiency. These challenges also confront the in-house development of large systems, where a response has been to focus on skill enhancement rather than automation (Ehn, 1989).

Challenges in Benefiting From User Contact

To benefit from verbal interaction, developers and users must learn one another's ways of talking about technology and its use or develop a new language.The importance of this difficult task is emphasized in recent Scandinavian projects (Bødker et al., in this volume; Greenbaum & Kyng, 1991). The use of prototypes can reduce the dependence on verbal communication, although the prototyping tool can introduce design constraints or fail to reflect future implementation constraints (Glushko, 1992).

Contact with users inevitably provides developers with only a partial understanding of the use situation. In particular, it is easier for developers to obtain insight into the nature of users' first encounters with a novel system, application, or feature; patterns of highly experienced use can be very important but are more difficult to explore prior to product completion. As a result, developers often just assume that heavy users will resemble themselves, diluting the effect of any user involvement.

For user contact to be translated into user involvement, it must have an impact on product design. When the identity of future users is uncertain and a wide range of conceivable candidates exists, a team may find it difficult to assess their experiences with a small number of possible users. The Scylla of overgeneralizing from a limited number of contacts is accompanied by the Charybdis of bogging down when users disagree. Evaluation is particularly challenging for systems that support groups (Grudin, in press). Finally, if user involvement does result in design recommendations, only the first step has been taken. Design ideas must be steered through a software development process that is typically fraught with obstacles to interface optimization (many of these obstacles are explored in Grudin, 1991b). User involvement can increase the odds of successfully navigating this course, but the journey is rarely easy, for reasons described later.

Challenges in Obtaining Feedback From Existing Users

Feedback from users may be collected informally or through bug reports and design change requests. The latter generally focus on what is of primary importance to customers—in particular, hardware reliability and high-level software functionality—and not on interface features. The little information that *is* relevant rarely gets back to developers. It is more likely to be available to those involved in subsequent product definition, although they may not be in a position to utilize it. Field service or software support groups shield developers from external contacts by maintaining products and working with customers on specific problems. The original project team members move on to new releases or product replacements, are reassigned to altogether different products, or leave the company for greener pastures.

The extent of feedback varies with the pattern of marketing and product use. A company with a heavy proportion of discretionary purchases initiated by actual users rather than by management or information systems specialists, such as Apple, *benefits* by having a particularly vocal user population. But in general, although user feedback is a very limited form of participation, the lack of it is among the greatest hindrances to good product interface design and among the least recognized defects of standard software development processes. System developers cannot spend *all* of their time fielding requests from customers, but their overall lack of exposure to feedback is an obstacle both to improving specific products and to building an awareness of the potential value of user participation in design. Developers rarely become aware of the users' pain.

This point deserves emphasis. Engineers are engaged in a continuous process of compromise, trading off among desirable alternatives. Interface improvements will be given more weight if engineers are aware of the far-reaching, lasting consequences of accepting an inferior design. Consider some typical tradeoffs: "This implementation will save 10K bytes but be a little less modular." "This design will run a little faster but take a month longer to complete." "This hardware configuration provides two more slots but adds $500 to the sales price." Each tradeoff requires a decision. Once the decision is made, the price in development time, memory size, or chip expense is paid and the matter is left behind. In this environment, the interface is just one consideration among many. "This interface would be nicer, but take two months longer to design." The decision could adversely affect thousands of users daily for the life of the product, but without feedback, the developer remains unaware of this special characteristic of the interface. Once it is built and shipped, they are on to the next job, and other people (including users) must do the sweeping up.

Trying to Find the Design Team

User involvement would be easier if one group had responsibility for all aspects of usability, as recommended by Gould (1988). But the "user interface," broadly defined, is not often the province of one recognizable team in a large product development company. The hardware is designed by one group, the software by another, the documentation by a third, and the training by a fourth. Representatives from other groups have periodic involvement—reviewing design specifications, for example. A product manager with little direct authority over developers may coordinate scheduling. Individuals from various marketing groups, such as competitive or strategic analysis[3] and international marketing, may contribute. Members of support groups such as human factors or performance analysis may participate, although not necessarily throughout the project. Several levels of management may monitor the process and comment at different stages. In concert, these people contribute to defining a computer user's experience with the system or application, yet communication among them may be surprisingly sparse. With whom are users to participate? In addition, turnover in project personnel is common, a further obstacle to sustained user involvement.

Matrix management is one approach to overcoming organizational separation. Representatives of a subset of these groups are given temporary assignment to a project. However, due to the perception or the reality that the contribution of "support roles" is limited to certain phases of a project, such assignments are often of limited duration or effectiveness. (Grønbæk et al., in this volume, describe a matrix management effort succumbing to these forces.)

The general neglect of on-line help illustrates divided responsibility affecting interface design. A good help system can save a company a substantial sum in customer "hand-holding," service calls, printed documentation, and so forth. The savings would be in the budget of, say, the Customer Service Department. But the effort and expense would have to come from Development, who may get more credit for devoting their resources to producing new functionality instead. Missing is "affirmative action" to promote on-line help in the face of a lack of developer empathy with less experienced users that is reinforced by the lack of mutual contact. Thus, help systems often end up with a low development priority.

---

[3]Competitive analysis may seem to be a logical ally of a development organization. However, in practice their concern is often the effective marketing of existing products against competition, possibly influencing product definition, but not having much contact with subsequent development.

The Software Development Process

This section turns from primarily structural aspects of product development organizations to consider the effects of some widely used software development processes and methods. These methods were developed when interactive systems were rare. Many were developed in the context of contract or internal development projects, not product development. As a result, they may obstruct rather than facilitate the development of usable interactive products.

Prior to the late 1970s, computing resources were too expensive to devote much to the interface. Nor was the demand great: Most computer users were engineers who understood the system or were willing to learn it. The computer use environment was similar to the computer development environment. Thus, the interface that developers constructed for their own purposes during development and debugging was often adequate of even *appropriate* for subsequent users, who were engineers or programmers engaged in similar tasks.

One legacy of this era is the persistence of the belief that the interface can be ignored or tidied up at the end of development. Late involvement in the software development process is a common complaint of members of support groups such as human factors and technical writing (Grudin & Poltrock, 1989). They are the project members most likely to advocate user involvement. If management is unaware of the need for *their* early and continual participation, how much support can be expected for *their* calls for early and continual *user* participation?

Over time, the original "engineering interfaces" are being replaced by interfaces developed for increasingly diverse user populations, following a pattern seen in other maturing technologies (Gentner & Grudin, 1990). Software development methods did not anticipate this change. New approaches to development that address these concerns are emerging (e.g., Boehm, 1988; Perlman, 1989), but have yet to be proven or widely adopted. One source of inertia is that insufficient information about user environments reaches developers: The degree to which development and use environments have diverged is not appreciated.

"Waterfall" models of software development arose in the context of large government projects. By their nature, competitively bid contracts emphasize written specifications: Contact with the eventual users may be forbidden or discouraged following the initial requirements definition, which occurs prior to the selection of the developers. Separate contracts may be written for system design, development, and maintenance. Approaches to development emerging from this tradition included structured analysis, where the task "establish man-machine interface" is relegated to one subphase of system development (De Marco, 1978), and Jackson System Development, which "excludes the whole area of human engineering in such matters as dialog design . . . it excludes procedures for system acceptance, installation, and cutover" (Jackson, 1983). Because such methods do not specify user involvement, project plans do not anticipate it. Development organizations are not structured to facilitate it and often work against it. This is obviously not ideal for interactive systems development, where early and continual user involvement has been an early and continually recommended principle for developing usable systems.

A "Catch 22" is that even *late* user involvement is blocked. Once the underlying software code is frozen, a fully functioning system is available for user testing—but at that very moment, documentation moves into the critical path toward product release. Since it is the software interface that is being documented, the interface is also frozen—before a user can try it out!

Prototyping and iterative design are recommended by the developers linked here to innovation in technique or methodology—Boehm, Ehn, Gould and Lewis, Perlman—and by many others. These go hand in hand—prototyping has little point if the design cannot be changed. Unfortunately, the high visibility of the interface works against iterative design in three ways: (a) the interface is grouped with aspects of the product that must be "signed off" on early in development; (b) support groups, such as those producing documentation, training, and marketing, are strongly tied to the software interface and are affected by changes; (c) iteration or change in the interface is noticed by everyone, which can create uneasiness, especially in a discipline that has a history of stressing careful initial design.

The emphasis on careful early design makes sense for noninteractive software, with its relatively predictable development course. It works less well for the interface, where design uncertainty is inevitable—the motivation for prototyping and iterative design in the first place. As noted earlier, as the interface grows in importance, it will join the proposed functionality in the product definition, the preliminary design. And once management has signed off on it, change requires approval. Poltrock (1989) observed the unique problems that high visibility and dependencies create for interface developers. One developer summed it up: "I think one of the biggest problems with user interface design is that if you do start iterating, it's obvious to people that you're iterating. Then people say, 'How is this going to end up.' They start to get worried as to whether you're actually going to deliver anything, and they get worried about the amount of work it's creating for them. And people like (those doing) documentation are screwed up by iterations. They can't write the books. Whereas software, you can iterate like mad underneath, and nobody will know the difference."

Interface development is distinct from other software development. Gould and Lewis (1985) summarize it this way:

> 'Getting it right the first time' plays a very different role in software design which does not involve user interfaces than it does in user interface design. This may explain, in part, the reluctance of designers to relinquish it as a fundamental aim. In the design of a compiler module, for example, the exact behavior of the code is or should be open to rational analysis. . . Good design in this context is highly analyt-

ic, and emphasizes careful planning. Designers know this. Adding a human interface to the system disrupts this picture fundamentally. A coprocessor of largely unpredictable behavior (i.e., a human user) has been added, and the system's algorithms have to mesh with it. There is no data sheet on this coprocessor, so one is forced to abandon the idea that one can design one's algorithms from first principles. An empirical approach is essential.

Solutions to the problems can be found—*will* be found—but the problems are new and adopting the solutions will require changing the way we work. Unfortunately, an innovative process proposal is unlikely to leave management as comfortable as a detailed product design specification.

The Routinization of Development

As competition and the pace of change increase, product development companies are pressured to turn out enhancements and new products in a timely, predictable fashion. Consider this analysis by Mace (1990): "Ashton-Tate's decline began with what is becoming a well-worn story in the industry: failure to upgrade a market-leading product. Dbase III Plus went for almost three years before being upgraded, while competitors' products were upgraded as often as twice in that time." A similar pattern of predictable new releases (accompanied by minor real innovation) is found in other maturing markets, from automobiles to stereo systems. The result is pressure for a predictable and controllable software development process: for routinization of development. Parker (1990) describes a perceived solution to the problem described in the previous quotation: "Lyons (an Ashton-Tate executive) responds that he can keep customers by providing predictable if not always exciting upgrades. 'Customers don't want to be embarrassed; they want their investment to be protected. If you are coming out with regular releases, even if they skip a release because a particular feature is missing, they will stay (with the product) because the cost of change is large.' "

This perceived need for controlled development creates difficulties for approaches that have uncertain duration or outcome. Interface design in general has a relatively high level of uncertainty and user participation can increase development time and introduce the possibility of changing its direction. This is the intent, of course—to produce a better product—but it nevertheless works against these powerful pressures for predictability. One way to reduce the uncertainty is to standardize design as much as possible; in the case of interface design, future development could well take place largely within the context of the current generation of "desktop metaphor" interfaces, thus greatly constraining the design space.[4]

---

[4]Arguments that discount an increased deskilling of programming have focused on the internal software development groups in large corporations (see Friedman, 1989), not on software product development; the latter is where the competitive pressures that motivate increased control are most evident.

Positive Conditions for User Involvement

We can conclude with some optimism by noting that these companies also provide support for involving users in design, primarily by putting the human-computer interface in the spotlight.

A better interface is one way to distinguish a product and to increase its acceptance in a competitive marketplace. Applications are reaching out to "discretionary users," people who have the choice of whether or not to use a computer. The greater availability of alternatives further increases buyer discretion, and computer users are likely to consider usability in making choices.

Large product development organizations often have considerable resources to devote to usability—development costs are highly amortized. As noted before, these companies are major employers of human factors engineers and interface specialists.

There is a positive side to the relatively frequent upgrades and product replacements: Developers can break out of "single-cycle" development. Evaluation of existing product use can feed into the design of later versions and good ideas arriving too late for use on a specific development project can be retained for later use (Grudin, Ehrlich, & Shriner, 1987).

Product development efforts may have a large supply of potential users; in contrast to internal development, the fate of a product doesn't depend heavily on situational factors that operate in any given site.

Finally, while inertia may develop, software product development companies were founded on change and recognize at a deep level that they must change to survive. This leads to some openness to experimentation.

OVERCOMING THE OBSTACLES

To one working within a large product development organization, the obstacles sometimes seem insurmountable. But as just noted, the company has a powerful incentive to improve product interfaces. Ease of learning and use becomes a more important marketing edge as software products mature. Adding a new bell or whistle may not help much if the already available functionality is under-utilized. In addition, declining hardware and software costs permit more resources to be directed to the interface. These forces have already pushed large product development companies into the forefront of human factors research and development. In the long term, organizational structures and development processes may evolve, institutionalizing solutions to the problems described here. The forces in development companies that work systematically *against* user involvement stand in the way of product optimization and success.

The directions that these companies will take are not obvious. As the focus of development shifts from generic products to systems and applications that meet the needs of different specific markets, companies must choose among working closely with independent developers, working with value-added resellers who in turn work with end-users, and working with the diverse computer users themselves. Each alternative will benefit from user participation—but the identity of the users is different in each case.

In the meantime, where can developers look for approaches to overcoming these obstacles? Persistence should not be underestimated. There are examples of successful case studies and general approaches, such as the development of the Olympic Message System at IBM (Gould et al., 1987) and the contextual interview approach at Digital described by Holtzblatt and Jones in this volume.

Experiments with prototype testing and iterative development are increasing our understanding of when and how they are most effectively used. Boehm's (1988) spiral model of development builds these techniques into a disciplined software engineering methodology. He is one of several writers encouraging an explicit change of focus in development from the current "product focus" to a focus on process *per se* in development. Grønbæk et al. (this volume) describe a project that succeeded only after this shift occurred in midcourse.

Product developers have demonstrated an interest in learning from European experiences at Computer-Supported Cooperative Work, CHI, and the Participatory Design Conferences. The in-house development context of the European projects removes or reduces some of the obstacles described in this chapter. In achieving user involvement, these projects have informed the issues and honed the techniques. Finding ways to transfer what is learned in in-house development to product development is itself a challenge. The UTOPIA project (described in Ehn's chapter in this volume) explicitly applied some of these approaches to a product development effort. A small set of potential users was heavily involved with the developers, while techniques such as a newsletter brought in a broader selection of potential system users on a more limited basis during design and development.

WORKING STRATEGICALLY

As noted earlier, product developers acquire information about product users in many ways. Their own management and marketing department are expected to serve this purpose. Friends in field service or training, application domain experts hired into the company, consultants, information system specialists in user organizations, user groups, standards organizations, professional and trade magazines, and other conduits exist. Limited forms of user involvement may already be practiced by human factors engineers, technical writers, or others. While each of these approaches can be criticized, sometimes they work. They must be analyzed and understood. How did they come into use? Are they adequate for a given situation today? Will they be adequate in a future of greater user diversity and rising user expectations? One reason to understand them is that introducing participatory design will change the organization and will affect the lives of people who are engaged in roles that serve indirect communication. The alternative methods by which product developers learn about users may relieve some of the pressure for participatory design. The advocates of these alternatives will resist change. A new approach—participatory design—may be better received on a project where a strong case can be made that the existing alternatives are inadequate.

Due to the growing demand for more usable systems, practitioners may find a climate for modest experimentation with participatory design. But even to *begin* working effectively requires a clear awareness of the obstacles, an understanding of why they are there, and a tolerant recognition that their source is in institutional constructs, not in unsympathetic individuals.

ACKNOWLEDGMENTS

This work profited immeasurably from interactions over the years with Susan Ehrlich Rudman and Steve Poltrock. Eric Roberts provided a list of the affiliations of the Participatory Design Conference registrants and Don Patterson provided CHI Conference registration statistics. MCC and its Shareholders supported some of the research, and Aarhus University provided two key ingredients: an external perspective from which to view the product development environment and the time to write about it.

REFERENCES

Boehm, B. (1988). A Spiral model of software development and enhancement. *IEEE Computer, 21*(5), 61–72.

Bødker, S. (1991). *Through the interface: A human activity approach to user interface design.* Hillsdale, NJ: Lawrence Erlbaum Associates.

De Marco, T. (1978). *Structured analysis and system specification.* New York: Yourdon.

Ehn, P. (1989). *Work oriented design of computer artifacts.* Hillsdale, NJ: Lawrence Erlbaum Associates.

Friedman, A. L. (1989). *Computer systems development: History, organization and implementation.* Chichester, UK: Wiley.

Gentner, D. R., & Grudin, J. (1990). Why good engineers (sometimes) Create Bad Interfaces. *Proceedings of the CHI'90 Conference of Human Factors in Computing Systems,* 277–282. New York: Association For Computing Machinery.

Glushko, R. J. (1992). Seven ways to make a hypertext project fail. *Technical Communication, 38,* 3.

Gould, J. D. (1988). How to design usable systems. In M. Helander (Ed.), *Handbook of human-computer interaction* (pp. 757–789). Amsterdam: North-Holland.

Gould, J. D., Boies, S. J., Levy, S., Richards, J. T., & Schoonard, J. (1987). The 1984 Olympic

Message System: A test of behavioral principles of system design. *Communications of the ACM, 30,* 758–769.

Gould, J. D., & Lewis, C. H. (1983). Designing for usability—key principles and what designers thing. *Proceedings of the CHI'83 Conference on Human Factors in Computing Systems,* 50–53. New York: Association For Computing Machinery.

Gould, J. D., & Lewis, C. (1985). Designing for usability: Key principles and what designers think. *Communications of the ACM, 28,* 300–311.

Greenbaum, J., & Kyng, M. (Eds.). (1991). *Design at work: Cooperative design of computer systems.* Hillsdale, NJ: Lawrence Erlbaum Associates.

Grudin, J. (1989). The case against user interface consistency. *Communications of the ACM, 32,* 1164–1173.

Grudin, J. (1990). Groupware and cooperative work: Problems and Prospects. In B. Laurel (Ed.), *The art of human-computer interface design* (pp. 171–185). Reading, MA: Addison-Wesley.

Grudin, J. (1991a). Interactive systems: Bridging the gaps between developers and users. *IEEE Computer, 24,*(4), 59–69.

Grudin, J. (1991b). Systematic sources of suboptimal interface design in large product development organizations. *Human-Computer Interaction, 6,*(2), 147–196.

Grudin, J. (in press). Groupware and social dynamics: Eight challenges for developers. *Communications of the ACM.*

Grudin, J., Ehrlich, S. F., & Shriner, R. (1987). Positioning human factors in the user interface development chain. *Proceedings of the CHI + GI'87 Conference on Human Factors in Computing Systems,* 125–131. New York: Association For Computing Machinery.

Grudin, J., & Poltrock, S. (1989). User inferface design in large corporations: Communication and coordination across disciplines. *Proceedings of the CHI'89 Conference on Human Factors in Computing Systems,"* 197–203. New York: Association For Computing Machinery.

Jackson, M. (1983). *System development.* Englewood Cliffs, NJ: Prentice-Hall.

Kapor, M. (1987, January). Face to face. *INC. Magazine,* pp. 31–38.

Mace, S. (1990). Defending the Dbase turf. *InfoWorld,* January 8, pp. 43–46.

Parker, R. (1990). Bill Lyons' task: Incremental moves to consistency. *InfoWorld,* January 8, p. 44.

Perlman, G. (1989). Asynchronous design/evaluation methods for hypertext technology development. *Hypertext'89 Proceedings,* 61–81. New York: Association For Computing Machinery.

Poltrock, S. E. (1989). Innovation in user interface development: Obstacles and opportunities. *Proceedings of the CHI'89 Conference on Human Factors in Computing Systems,* 191–195. New York: Association For Computing Machinery.

Poltrock, S. E., & Grudin, J. (preparation). Participant-observer studies of interface design and development.

III APPROACHES TO PARTICIPATORY DESIGN

7

Ethnographic Field Methods and Their Relation to Design

Jeanette Blomberg
Xerox Palo Alto Research Center

Jean Giacomi
Andrea Mosher
Pat Swenton-Wall
Xerox Corporation[1]

In this chapter we explore the relationship between developing a descriptive understanding of human behavior and designing artifacts which ostensibly support the activities described. Although there is growing recognition that an understanding of users' current work practices would be useful in the design of new technologies, the debate about what it would mean to acquire such understanding and to link it with design is only beginning. What are the implications of developing ways of representing the views and activities of communities of practice outside one's own such that the knowledge would be useful in design?

The ethnographic approach, with its emphasis on "natives' point-of-view," holism, and natural settings, provides a unique perspective to bring to bear on understanding users' work activities. However, anthropology is mute when it comes to ways of integrating such an understanding with design. The languages of design and of ethnography evolved in quite different contexts and in relation to different concerns. While the ethnographer is interested in *understanding* human behavior as it is reflected in the lifeways of diverse communities of people, the designer is interested in *designing* artifacts that will support the activities of these communities. The current challenge is to develop ways of linking these two undertakings.

[1]We wish to thank members of the Industrial Design/Human Interface Participatory Design Project who worked with us in our exploration of the relevance and power of ethnographic field methods for design.

As practiced by most ethnographers, developing an understanding of human behavior requires a period of field work where the ethnographer becomes immersed in the activities of the people studied. Typically, field work involves some combination of observation, informal interviewing, and participation in the ongoing events of the community. Through extensive contact with the people studied ethnographers develop a descriptive understanding of the observed behaviors.

Designers, on the other hand, are interested in understanding human behavior insofar as it enables them to *design* artifacts better suited to the needs of the users. Designers, therefore, spend more time testing and evaluating their designs in relation to users' needs and abilities and less on understanding the supported behavior per se. When designers do attempt to gain a clearer view of the users for whom they design technologies, they traditionally have been limited in the ways such a view is acquired (see section on Traditional Approaches, pp. 143–147). Ethnography provides an alternative methodology for designers to use, which gives them access to people's *everyday practices* as members of social groups.

MOTIVATION FOR USING AN ETHNOGRAPHIC APPROACH

In the 1980s there was a refocusing of interest on the part of systems designers away from the view that technology supported individual tasks and toward the view that human activities were in large part carried out in cooperation with others. Therefore new technologies should be designed to support the cooperative nature of most human activities. A new field evolved called CSCW (computer support for cooperative work), which was concerned with the design of computer tools for the support of group work (Greif, 1988). As a consequence of this shift in focus, there was a realization that the methods most often used to analyze users' needs and activities, and to evaluate designs, were not suited to this changed focus. Looking at individual psychological and cognitive processes and evaluating the fit between isolated tasks, users, and technologies would not provide the perspective needed to design and evaluate technologies for group work. Interest in exploring the possible applicability of ethnographic methods for understanding group work practices and for linking this understanding to design coincided with, and may have been motivated by, this refocusing of interest.

THE ETHNOGRAPHIC APPROACH

There is considerable debate in anthropology about what it means to be "doing" ethnography. At a minimum, most would agree that ethnography requires a period of field work where the ethnographer becomes involved in the everyday activities of the people studied. While ethnography often includes a description of the activities and practices of those studied, it is more importantly an attempt to interpret and give meaning to those activities. This interpretation most often takes its final form as a text written in a somewhat stylized format (see Van Mannen 1988, for a discussion of various ethnographic forms). Although many books have been written about ethnography and ethnographic field methods, there is no agreed upon set of principles that guide anthropological field work, nor is there a cookbook of methods and techniques applicable in all situations. We offer here some general guidelines for those interested in exploring the usefulness of an ethnographic approach for the design of new technologies and present a brief description of some commonly employed research methods. However, we would falsely portray the field if we left the impression that what we write here is either complete or without challenge. We have chosen not to dwell on the controversies that exist concerning what constitutes an adequate ethnography to avoid complicating the discussion. Those interested in delving further into the controversies should consult Agar (1980, 1986), Clifford (1988), Clifford and Marcus (1986), Geertz (1973, 1983), Harris (1979), Naroll and Cohen (1970), Pelto (1970), Tyler (1979), and Van Mannen (1988), among others.

Guiding Principles of Ethnography

At the risk of overly simplifying the ethnographic endeavor, we begin by describing four main principles that guide much ethnographic work.

Natural Settings

Ethnography is grounded in field work. By this we mean that there is a commitment to study the activities of people in their everyday settings. This requires that the research be conducted in a field setting as opposed to a laboratory or experimental setting (see Fig. 7.1). The underlying assumption here is that to learn about a world you don't understand you must encounter it firsthand.

Holism

This emphasis on natural settings derives in part from a belief that particular behaviors can only be understood in the everyday *context* in which they occur. To remove a behavior from the larger social context is to change it in important, nontrivial ways. This concern with how particular behaviors fit into the larger whole is often referred to as holism (see Fig. 7.1).

Descriptive

Based on field work ethnographers develop a *descriptive* understanding of the lifeways of the group studied. Ethnographers describe how people *actually* behave, not how they *ought* to behave. This distinction is similar to one made in

HOLISTIC

Particular behaviors understood in relation to how
they are embedded in the social and historical fabric
of everyday life.

Focus on relationship between the parts

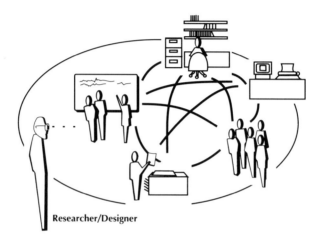

Researcher/Designer

Learn about a world you don't understand by
encountering it first hand. Focus on naturally
occurring, everyday talk and action.

NATURAL LABORATORY

FIG. 7.1.   Two principles of ethnograpy.

linguistics between descriptive linguists (how people speak) and prescriptive linguistics (how people ought to speak). The orientation toward the descriptive leads ethnographers to assume a nonjudgemental stance with respect to the behaviors they study. Maintaining such a nonjudgemental stance is sometimes referred to as cultural relativism, the notion that other people's behaviors should not be judged by the standards of some other group (see Fig. 7.2 for an illustration of how descriptive and prescriptive characterizations contrast.)

DESCRIPTIVE

Judgements of the efficacy of
behaviors observed are
withheld

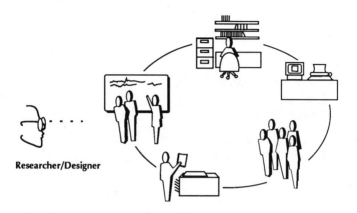

Researcher/Designer

DESCRIPTIVE——————— *Instead of* PRESCRIPTIVE ———————

"Several people handle the
document before it is completed.
All involved discover problems
and are asked to account for
changes to the document."

"They're still manually
processing these "routine"
documents. Passing hardcopy
from person to person is such
an inefficient way to update
documents. An electronic mail
system linked to an intelligent
database could really improve
their process."

FIG. 7.2. Contrast between descriptive and prescriptive characterizations of activity.

Members' Point-of-View

Ethnography involves understanding the world from the point-of-view of
those studied. Anthropologists attempt to understand how people organize their
behavior and make sense of the world around them. With the realization that one
can never truly get inside the head of another or see the world exactly as another
does, research methods are aimed at getting as close to an insider's view of the
situation as possible. With such an orientation, ethnographers are concerned with

describing behavior in terms relevant and meaningful to study participants. This contrasts with the requirements of survey research where relevant categories must be known before the study begins and must not vary across participant communities. As such, the terms in which behavior is characterized often are those of the researcher, not the study participants (see Figs. 7.3a and 7.3b for an illustration of this principle).

In general then, ethnography is concerned with understanding other people's

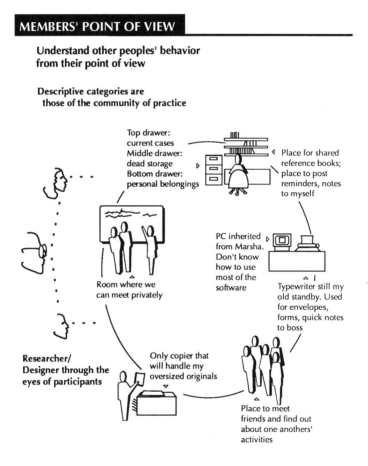

FIG. 7.3a. Descriptive categories of the study participants as contrasted with those of the researcher (Fig. 7.3b).

Contrasted With _____

Descriptive categories are
those of the researcher

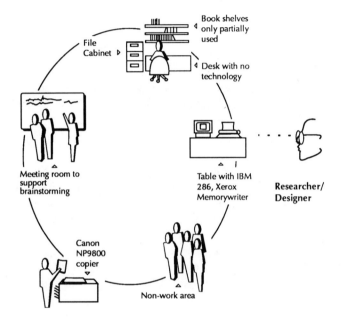

FIG. 7.3b. Descriptive categories of the researcher as contrasted with those of the study participants (Fig. 7.3a).

behavior in the context in which it occurs and from the point-of-view of the people studied.

ETHNOGRAPHIC FIELD METHODS

Ethnographic field work requires the personal involvement of the investigator, a willingness to be in situations out of one's control and as such an abandonment of strict "scientific control." It also involves an iterative, improvizational approach to understanding, wherein partial and tentative formulations are revised as new observations challenge the old, and where adjustments in research strategy are made as more is learned about the particular situation at hand.

Although ethnography cannot be reduced to a set of methods or techniques, but must be understood in relation to the above principles, we discuss a few commonly employed field techniques that have proven useful to anthropologists in the field.

Observation

Because ethnographers are interested in understanding human activity in the everyday settings in which it occurs, most ethnographic investigations involve some period of observation. The ability to observe and record ongoing activities becomes critical to the success of the endeavor.

Why Observe?

There is a well-known axiom in anthropology that what people say and what they do are not the same. This is one of the principal motivations for including observations of ongoing activity in any study of human behavior. The distinction between what people say and what they do is related to the distinction between *ideal* and *manifest* behavior. Ideal behavior is what every "good" member of the community should do, whereas manifest behavior is what people actually do. Sometimes asking people about their behavior will yield responses closer to the ideal than the manifest. People may distort, either knowingly or unknowingly, accounts of their own behavior, often simply providing an approximation constructed either for the questioner's benefit or to match cultural expectations. With regard to this latter point, Whiting and Whiting (1970) give examples from studies of child rearing where women's descriptions of their child rearing practices were at odds with their observed practices. Women in the United States reported that their children could dress and toilet themselves earlier than they actually could. Whiting and Whiting assert that this supported the value these mothers placed on independence.

The distinction between what people do and what they say is also related to the fact that people often don't have access to the inarticulated, tacit knowledge associated with certain activities. There are many activities that are so much a part of our everyday lives that we are unable to provide accurate accounts even when asked to reflect upon these activities. In some cases, we may not have the vocabulary to talk about them.

Observational Role

There are many ways to conduct observations. At one extreme one may attempt to become the proverbial "fly on the wall," in which case the field worker tries to be as *unobtrusive* as possible. Maintaining the strictly observer role is difficult and frequently requires being given some culturally appropriate role that allows the observer to "hang around" and observe.[2] Some have characterized this role as that of the *observer participant,* where the participant component is simply the culturally appropriate status given to the ethnographer.

At the other extreme is the *participant observer.* In this case the field worker becomes a full participant in the activities studied[3] and has the opportunity to be an observer as well. There are both advantages and liabilities in attempting to be both observer and participant. One major advantage is that the field worker has firsthand experience of the events under study due to the fact of active participation. In some settings the only way of gaining access to the activities of the community is by assuming an active role; simply being an observer to events is not acceptable (Kluckhohn, 1940).

On the other hand, the liabilities of being both observer and participant are many. Logistics alone can be overwhelming if one is trying to participate and take field notes at the same time. It often becomes necessary to write-up field notes as after the fact recollections of events participated in. Because the ethnographer is often new to the activities participated in, a great deal of energy can be expended simply figuring out what a next appropriate action might be. Under these conditions participating at the same time you are trying to make sense of the totality of events as they unfold may be impossible.

As a participant one has both the advantage and disadvantage of having a view of the activities from a particular vantage point, one's participatory role. Participating in a different capacity would provide a somewhat different perspective on the events. While this may seem like a problem, every account provides a view from somewhere, if not from a participant's standpoint, then from the outside and some particular outside at that. There are always multiple vantage points from which to gain an understanding of the activities of a community and each will enlighten in different ways.

The unobtrusive and participant observer roles are two extremes along some continuum. Most often one moves back and forth between participation and observation, variously wearing the hat of the insider and outsider. Different studies and observational episodes within the same study provide different opportunities to either take an unobtrusive observer role or to participate in the ongoing activities of the community. One should not feel bound to choose among these observational extremes.

Observations of whatever variety require that the field worker maintain good

<hr/>

[2]While unobtrusive observation may be desirable, there are ethical issues that surround observing or videotaping people for study without their awareness. To the greatest extent possible, study participants should be informed about observations or videotaping of their activities.

[3]Full participation in the activities of the community under study has some advantages. However, anthropologists have cautioned fieldworkers that "going native" might have negative consequence both for the research project and for the well being of the investigator. This is a greater problem when field work is conducted far from home, among markedly different cultural groups.

relations with the people observed. The field worker must be socially skilled and aware of the sensitivities of those whose activities are being observed. Jeopardizing one's position in the community by insisting on observing or participating in an activity, when doing so is deemed inappropriate by the community, could easily spell disaster for the project. It is more advantageous to proceed slowly, gaining trust before insisting on access to certain events, people, and activities. Often patience and greater familiarity with local custom will pay off. What was off limits initially may become open in time.

Focus of Observation

Once the decision to observe is made, there are still many questions to answer. One must decide what to observe, when to observe, where to observe, and when you've observed enough (Whiting & Whiting, 1970). Answers to these and other questions will depend on research direction. One might decide to observe meetings (event focus), or individuals as they go through their daily routine (person focus), or the activities at and around a receptionist's desk (place focus), or the life history of a document as it moves from office to office and person to person (object focus) (see Suchman & Trigg, 1991, for a slight variation). Choices like these become necessary because of the impossibility of taking in all that is going on in a particular setting and because different research questions require different observational strategies.

Although there is no fixed rule concerning when one has observed enough, a general principle is that when you're no longer surprised by what you're observing, you've probably seen enough. In other words, when you can *predict* what will occur during some period and these predictions are consistently born out by repeated observation, you can be secure in believing that a range of behaviors and activities have been adequately sampled. However, this confidence depends on appropriate sampling strategies for the observational periods. For example, if one's interest is in the activities at or around a particular location, it is important to observe those activities at various times of the day. The morning activities around a receptionist's desk may differ significantly from those at the close of the day. The issue of sampling strategy is important no matter what the focus of observation (person, event, place, or object).

Note Taking

Note taking is a very individual activity, but it is one of the important links between the field experience and how one later interprets that experience (Jackson, 1990). Although field notes are never a complete record of the experiences and observations of the field worker, they are often used to evoke memories of experienced events. As such, field notes are most useful to the original field worker.

There are many strategies that have been developed for taking field notes.

Different situations and different observational roles will influence the possibilities for taking field notes. As mentioned earlier, the full participant role may
make it impossible to take notes while events are unfolding. Some situations may
dictate that only sketchy, temporary notes can be taken at the time of observation,
the intent being to expand the notes soon after the event is over. Other situations
may allow taking very extensive notes which might include verbatim transcripts,
paraphrases of things said, interpretations of activities observed, as well as
descriptions, maps, or diagrams of the setting or the movements of people and
artifacts within the setting. When field notes include different types of information, it is often important to indicate the status of the information. Is it a verbatim
transcript or simply a rough paraphrase; is it a member's stated motivation for
participating in the event or a conjecture on the part of the ethnographer? While
these distinctions may seem clear enough at the time the notes are taken, one
must remember that field notes may be referenced months or years after they
were taken. In this light it is critical that field notes include at a minimum a date,
time, place, and listing of the persons present.

Videotaped Records as Notes

Video cameras have come to play an increasingly important role in ethnographic studies (see section on Video Analysis). They are sometimes used as a
supplement or even substitute to field notes. When memory fails or field notes
are inadequate, the videotape may be able to provide some of the missing
information. However, caution needs to be exercised when relying on the video
record for a complete record of observed events. Viewing a videotape at a later
date and experiencing an event firsthand provide the researcher with different
kinds of access to the activities in question. While the camera records those
activities within its field of view, participating in the activity allows one to absorb
the "taste, smell, and feel" of the activity, and to refocus attention in response to
unfolding events. A videotape cannot capture the ways the event is experienced
by the observer/participant.

Supplementing videotaped records with field notes also can help with later
analysis of the videotapes. Anyone who has recorded hours of videotape knows
how laborious it is log the tapes for content. Careful, painstaking analysis often
is conducted on only a very small portion of the available videotape. Field notes
can be a great help in selecting segments of the tape for more focused analysis.

Observations Coupled With Interviews

Observations seldom stand alone and are frequently coupled with interviews
and informal discussions. Because ethnographers make no assumption that appropriate questions or ways of asking them are known in advance, observations
provide ethnographers with one way to learn how to ask appropriate questions
from the point-of-view of the members of the community under study. Observa-

tions and informal discussions also follow interviews providing the opportunity to observe behaviors previously described by interview respondents.

Interviewing

Informal interviewing is a part of most ethnographic research. Early in field work these interviews are intentionally unstructured and open-ended to allow the participants to help shape the discourse, the topics to be discussed, and the relevant ways of talking about them. There are few assumptions going in about what questions to ask or how best to ask them.[4] The ethnographer may have general areas of interest to explore, but if the discussion moves away from these areas to issues more "relevant" to the respondent, this is likely to be viewed as an opportunity to learn about the community rather than a situation to be avoided lest field work proceed too slowly. Through participation in community activities and through informal interviews the ethnographer begins to learn enough about the community to conduct more structured, systematic interviews. The motivation to begin with informal interviews, as opposed to structured ones, derives from a belief that asking highly constrained questions before enough is understood about the situation likely will produce poor quality answers. In such a situation questions may be understood in unknown ways or may be irrelevant to the respondent, in which case the researcher might get a mistaken sense of the relationship between the questions asked and the responses given.

It also should be noted that interviewing, whether formal or informal, is not a simple way of recording objective fact (Mishler, 1986). As mentioned in the preceding section on observation, what people say and what they do are not the same thing. Asking people to describe some activity in which they engage will not produce the same insights as one would gain by observing people engaged in the activity. In the same way asking people about their beliefs will not tell you about how those beliefs are manifest in their everyday activities. If we simply relied on the interview as our window to some objective reality, we might come to some very erroneous conclusion about the lifeways of the communities we study.

If instead we view the interview (following Briggs, 1986) as a communicative event, we must know something of the interactional dynamics that shaped the interview to make sense of the discourse (the questions and responses). As Briggs (1986) states, "Like speech events in general, it (the social situation created by the interview) shapes the form and content of what is said" (p. 22). The implica-

---

[4]When field work is conducted among people who speak a different language from the ethnographer, a point is made to learn their language so that questions can be asked in the language and vernacular of the study participants.

tions of this are twofold. First, the context[5] in which the interview takes place is not neutral with respect to the discourse that ensues. Second, to make sense of what is said during an interview one needs to examine the social construction of the questions and responses in relation to the social situation created by the interview itself (Suchman & Jordan, 1990).

Interview Location

Because ethnography is field-work based, interviews most often occur in the local setting. There are advantages to interviewing in the respondent's environment. Not only are the respondents more likely to feel comfortable in familiar surroundings, but they have access to people and objects that may figure into the talk as it unfolds. If a respondent is trying to describe an activity in which he participates, having available the artifacts, physical surroundings, and people that typically help shape the activity can be a resource for the talk. This said one must also be aware of situations where the respondent's environment does *not* provide the privacy needed to talk about some subjects. If other members of the community can overhear, respondents may restrict what they are willing to talk about. The point here is that the setting for the interview is never neutral with respect to how the interview unfolds and this must be taken into consideration in any subsequent analysis of the responses.

Contextual Interviewing

Interviewing can be combined with observation where the researcher interviews respondents while they are engaged in some activity. In this situation the researcher "interrupts" the observed activity to ask questions, motivated by the observations being made. However, there is a trade off here. By asking a question about some activity, in the context in which the activity is taking place, one will influence the course of the activity. Bearing this in mind, the contextual interview can provide access to information that observation alone might fail to uncover.

Who to Interview?

Because it often is impossible to interview all members of a community, decisions must be made concerning who to interview. These choices are influenced by the research questions being posed, the availability of respondents, the

---

[5]The context is not simply the sum total of the physical and social characteristics of the situation. The context is continually being jointly produced by the participants and as such cannot be easily separated from the activities and talk in question. See Cicourel (1982) and Mehan (1979) for a discussion of these issues.

project time frame, and a concern with sampling the views and activities of a cross-section of the community. This issue of representativeness was not always a concern for ethnographers. Early ethnographers were not particularly concerned to find themselves in situations where they were relying on information elicited from one individual because they held the view that any member of a community could provide information about the beliefs, values, and customs of the entire community. However, as ethnographers began to question assumptions about cultural sharing, the practice of relying on a single informant (ethnographers' traditional term for respondent) all but disappeared. Ethnographers questioned whether the chosen informant (often a "westernized" or marginal member of the community or one who shared characteristics with the researcher) could adequately represent the beliefs and practices of all community members. Relying on a single informant became known in anthropological circles as the "well-informed informant problem."

Rules of Thumb in Interviewing

The ethnographic interview is not bound by explicit rules. To the contrary, a great deal of latitude is given to the interviewer to exploit the particulars of any given interview situation (see Suchman and Jordan, 1990 for a discussion of the pitfalls of rigid interviewer guidelines). Nevertheless, there are still a few basic rules, or what might be better described as interactional style considerations, that should be mentioned. First, it is important to allow respondents to help shape the content and character of the interaction. In keeping with this precept, it is generally not advisable to interrupt unnecessarily, to complete the respondents' utterances, or to answer your own questions. While the interviewer may be able to anticipate what a respondent is going to say, much more is learned by allowing the respondent to answer the question unaided by the interviewers' presuppositions. If the interviewer is mistaken about the answer contemplated by the respondent, there is the risk that the respondent will accept the interviewer's response, thinking it "sounds better" than the one contemplated or will not want to contradict the interviewer. In any event, an opportunity to learn more about the respondent's world will be lost.

A second guideline in ethnographic interviewing is that rapport with the respondent should not be sacrificed to obtain a response. The respondent may choose not to answer a question for any number of reasons (e.g., too personally revealing, concern that the information will get to other parties, inability to understand why the interviewer wants to know such information, etc.). Ethnographers rely on being able to find new, more appropriate ways of asking questions as they learn more about the setting and as the respondents become more familiar with them.

Third, it is important that the interviewer be willing to acknowledge knowing less than the respondent. After all, the point of the interview is to learn something about what the *respondent* knows or how the *respondent* sees the world. Although the interviewer may be more knowledgeable about some topic, the objective of the interview is *not* to produce the most accurate or complete understanding of the topic, but to gain a better understanding of what the respondent knows and thinks about the topic. The purpose of the interview may be defeated if the interviewer is more concerned with self-aggrandizement than listening to what the respondent has to say. A related problem when ethnography is part of a design project is the tendency of some designers, whose job after all is to solve problems, to come up with solutions to a respondent's expressed problem before taking the time to get an adequate understanding of the problem. The push to solve design problems may conflict with taking the time to fully appreciate the nature of the problem.

Video Analysis

The use of video cameras in ethnographic research is on the increase with the growing availability of inexpensive, small, portable equipment. There are a variety of ways ethnographers make use of video records of activity. For some they are a supplement to field notes, for others they are used in teaching and reporting situations, and for still others they are the primary data for analysis.[6] Space limitations preclude a detailed discussion of the techniques used in the analysis of video records, but we offer a few general arguments for bringing video cameras to the field.

Why Videotape?

One of the strongest arguments in favor of videotaping is that human activities unfold so fast that it is impossible to capture their complexity by observation alone (Jordan, Henderson, & Tatar, in preparation). Field notes are only a partial record of activities observed or participated in, and words are often inadequate to describe what is observed, including bodily movements relevant to the analysis. The videotape preserves these actions for careful viewing and analysis. Videotaped records also allow one to look at an activity from different perspectives from the one held at the time the video record was made.

As the field work progresses and the researcher develops new understandings of the activities, new perspectives can be brought to bear on the activities previously recorded. The ability to review videotapes also allows the researcher the opportunity to correct erroneous characterizations and interpretations (Suchman & Trigg, 1990). In this sense the researcher can corroborate the field record (notes) with the video record.

---

[6]Many ethnographers employ analytic techniques developed in the areas of conversational and interactional analysis. Those interested in learning more about these techniques and the assumptions that underlie them should refer to Atkinson and Heritage (1984) and Goodwin and Heritage (1990).

Another advantage of having a video record is that it can be made available to people not present at the time the recording was made. Field notes, absent an accompanying elaboration by the person who took them, are of limited use to anyone but the original note taker. On the other hand, videotapes of activity can be viewed and analyzed by a wide range of people (other researchers, designers, product developers, etc.). The participation in the analysis by people involved in the original videotaping often can facilitate the analysis and contribute to what can be learned from the tape.

Videotaping ongoing activity also provides another example of *unobtrusive observation* since a camera can be set up and the researchers can leave the scene.[7] Despite the opportunity to observe without influencing the course of events, in some circumstances it may not be desirable for the researcher to leave the scene. There are times when the investigator's presence is appropriate; when the investigator wants to be a participant in the recorded activity or when the investigator wants to be present to ask for clarification and elaboration. The investigator's presence in the scene should not determine whether videotaping is appropriate, but an awareness of the possible influence of the investigator's presence on the activities should be taken into consideration in subsequent analysis.

Problems With Videotaping

Videotaping is not problem free. Because one can quickly generate large quantities of tape all of which cannot be analyzed in detail, the researcher faces the problem of identifying sections of the tapes for careful, painstaking analysis. This is facilitated by logging the tapes soon after they are made, annotating them with general descriptions of activities and highlighting places where particular research questions are addressed. Field notes can be useful in producing these content logs, but if one was not present when the tape was made, the content logging activity will require viewing the entire tape at least once to produce an adequate content log. This can amount to hours of work. Once the content log is completed, particular sections of the tape can be selected for later careful analysis.

Analyzing videotapes is a time-consuming activity that cannot be delegated to others. Unlike survey research where interviewing participants, coding responses, and conducting statistical analysis can be done by others, the job of analyzing videotapes by repeated viewing can be done only by those who will be engaged in interpretation. Insights come only through participating in careful analysis (Heritage & Atkinson, 1984; Suchman & Trigg, 1990).

---

[7]While the influence of the camera's presence on behavior cannot be ruled out, in most cases the camera quickly becomes part of the background and only occasionally surfaces in the participants' awareness.

Another problem with the use of video cameras is that some human activities are difficult to record on videotape. There are times when a space may be too small to get an acceptable view with the camera, or when noise levels are not adequate for suitable audio recording. The spatial distribution of an activity may require multiple cameras to adequately record the activity and the changing location of an activity may require mobile cameras. Multiple (mobile) cameras are not possible in all settings and one may have to settle for recording only part of the activity. If it is possible to use more than one camera, new analytic challenges emerge. While the use of split screen images, time code to synchronize multiple recording, and computer controlled editing equipment makes such analysis possible, it is difficult nonetheless.

UNDERSTANDING HUMAN BEHAVIORS AS A MECHANISM FOR CHANGE

As mentioned, ethnography is a way of developing a descriptive understanding of human activities. Insofar as such an understanding can be brought to bear on designing new technologies, its role as a mechanism for change must be considered. To greater and lesser degrees new technologies always result in change for the communities into which they are introduced. As Ehn (1988) puts it, "What we design is not just artifacts but by intervention a changed or reformed practice" (p. 128). As such those involved in linking ethnography and design must be aware of their role as "change agents." This raises the question, as it does for anthropologists who act as change agents in more traditional settings; "In whose interest does one operate?" Does one serve the people for whom new technologies are designed (those whose activities are the subject of the inquiry) or does one serve the sponsors of the work? Arensberg and Niehoff (1971) contend the main concern of the anthropologist involved in promoting change ". . . must be with the people who he hopes will accept the new ideas" (p. 7). In many cases these are the end-users of the new technologies.

Because ethnography typically involves extensive contact with the people studied and an attempt to "see" the world through their eyes, ethnographers frequently identify with the interests of those studied regardless of research sponsorship. Van Mannen (1988) writes, ". . . the fieldworker not only represents but takes the side of the studied" (p. 42). This orientation toward the concerns of the people studied is a central characteristic of anthropology and has been codified in the "Principles of Professional Responsibility" adopted by the American Anthropological Association in 1971. The first principle states, "In research, an anthropologist's paramount responsibility is to those he studies. When there is a conflict of interest, these individuals must come first." (Ethics and Anthropology: Dilemmas in Field work, p. 183). As anthropologists become more involved in systems design and development, and as ethnographic field methods become more widely used in such efforts, it is important for those involved in this work to reflect upon these ethical concerns. In particular, those whose work is supported by technology companies must ask if it is possible to be "both user advocates and purveyors of technology?" (Kathleen Carter, personal communication).

PARTICIPATION IN FORMULATING DESCRIPTIONS OF NATIVE PRACTICES

Little has been written about the role of those studied in formulating descriptions of their own practices. While the ethnographer may attempt to represent the experiences of those studied from the "natives' point-of-view," active participation of the studied in reviewing and analyzing the ethnographer's formulations is rare. At most, the studied may become involved in checking the "accuracy" of aspects of the ethnographers field notes. The ethnographer's accounts and representations of native practices often are developed far from the field where participation of those studied is rarely possible even if desirable. In defense of the practice of not involving the studied more directly, some have argued that those studied are not in a good position to reflect on their own behaviors and would most likely simply offer post hoc rationalizations or justifications for their behaviors if asked to participate in analysis.

Ethnographers become concerned with issues of participation when they become involved directly in making recommendations for changes to native practices. When they take the role of change agents and their formulations could have material effects on those studied, participation becomes an important issue. Niehoff (1966), writes in a casebook of social change,

> No project which will effect socioeconomic change can possibly succeed if the recipients do not participate. . . . It is surprising how frequently action agents have neglected to make sure they have committed participation on their projects. It appears that the principal reason why this primary ingredient has been overlooked so often is that many technical advisers have viewed their task as simply one of providing some kind of technical solution (p. 18).

Anthropologists often are brought into action-oriented projects when there is the realization that the change agents need a better understanding of the practices and beliefs of the "beneficiaries" of the project and that these individuals need to be involved in the specification and integration of the new practices and technologies. The anthropologist become a link to these indigenous communities.

In attempting to link ethnography to design, it is important that ways of involving those studied in the specification of the new technologies and practices be developed. This is needed for several reasons: First, by involving those studied it is possible to gain new understandings of the studied activities in relation to the evolving design. Second, insofar as researchers and designers operate as "change agents," they need to respect the interests of those studied in whatever technology solutions are developed. Third, in those cases where those studied will become the actual users of the new technology, their early involvement may aid in adoption later on.

EXPECTATIONS OF THOSE PARTICIPATING IN AN ETHNOGRAPHIC STUDY

Ethnographers typically promise little in return for the opportunity to study a community of people other than to avoid major disruptions in the community's activities and to look after the community's interests in subsequent interactions with outsiders (e.g., publications, presentations, discussions). The argument has been made that ethnography is exploitive to the degree that it is not concerned with the *use* of the knowledge gained to better the conditions of the study participants. Strathern (1987) writes that people may experience exploitation ". . . when [they] perceive that others have the power to turn data into materials whose value cannot be shared or yielded back to them in return" (p. 20). Some have argued that one reason ethnographers often study the less advantaged is because these communities are not in a position to demand something in return for participation in the study. Nader (1974) states, "Anthropologists might indeed ask themselves whether the entirety of field work does not depend upon a certain power relationship in favor of the anthropologist, and whether indeed such dominant-subordinate relationships may not be affecting the kinds of theories we are weaving" (p. 289). Issues of access become salient as ethnographers attempt to study franchised and more powerful communities.

When ethnography is a part of a technology development effort issues of access and reciprocity must be confronted. In some situations members of franchised communities must be convinced to allow access to the settings in which they work, without the promise of providing them with a technology solution. This may be the case because the technology under development may never become commercially available or, if it does, it might be years before it is on the market. The ability to gain access to the communities of study and the promises that can be made about materially bettering the lot of those studied is directly linked to the type of technology development effort undertaken.

WHY IS ETHNOGRAPHY RELEVANT TO DESIGN?

Ethnography is relevant to design for several reasons. First, since designers often create artifacts for work settings they know little about, some understanding of those settings is needed so that the technologies suit the situations of their use.

Second, because technologies help shape the work practices of their users (Ehn, 1988; Blomberg, 1987, 1988a), it is important that the designers' world view not be imposed inappropriately on users. If designers have little information regarding the situations in which technologies are used, the best they can do is rely on their own experiences and imagination thus running the risk of designing technologies better suited to their needs than those of the actual users.

Third, there are situations where designers create technologies whose possible uses are unknown. Such situations might be described as technology in search of an application. Some understanding of the work in which potential users are engaged can help identify possible uses and refine the original technology design.

Fourth, since the user's experience of a technology is influenced by the *context* of its use (Blomberg, 1987, 1988a), gaining a broader perspective on technology use than that arrived at through traditional operability testing (with its focus on the human-machine dyad) is important (see section on Traditional Approaches, p. 144, for a description of operability tests).

Fifth, when designing radically new technologies, users often are unable to give meaningful responses to queries about how they might use such technologies. They need to be provided with a way of *envisioning* and experiencing the technology in the context of their own work practices before they can contribute to such a discussion. To create the context for such a discussion and to be useful partners in the joint exploration of the relation between work and technology, designers must have some understanding of the user's work (Blomberg & Henderson, 1990).

Finally, the single-task focus of some technology design efforts is ill-suited to the design of technologies that support task *integration*. Simply focusing on a single task or the tasks of the single user ignores how the work of one individual articulates with that of many others. For example, a print shop operator's work may rely on the work of document creators, word processing specialists, graphic artists, sales representatives and many others. Systems that support the print shop operator's work should be designed with some larger understanding of how the work of these others impinges upon the work of the print shop operator.

LINKING ETHNOGRAPHY AND DESIGN

There are various ways one might imagine acquiring, representing, and transferring the knowledge gained from an ethnographic analysis of user work practices in the context of technology design. We will mention only a few of them. First, a trained ethnographer might be asked to study the work practices of some group. The insights from this study might then be transferred to designers through written reports and oral presentations. The *designers* would then have the task of identifying the relevant aspects of the reports for their particular design efforts.

Because of differences between the languages and perspectives of ethnography and design and because the ethnographer is likely to have little knowledge or appreciation for the immediate concerns of the designers, this is not a simple task. Making the findings of an ethnographic study useful for day-to-day design concerns becomes a major undertaking.

Second, an ethnographic study might be undertaken by a team of investigators consisting of ethnographers and designers. In this case the insights and under-standings, in part, would be embodied in the experiences of the *designers* who were firsthand participants in the study. As Penniman (1974) observed, experi-ence underlies all understanding of social life. Active involvement by designers in the field work and in constructing interpretations of the work activities at the study site also would help focus the ethnographic study on issues more central to the design task and would make the interpretations more relevant to the design.

Third, a project could be undertaken by a team of ethnographers, designers, and *users*. The understandings and insights derived from the study would not necessarily be represented in a written report, but instead would be reflected in a codesigned artifact. User partnership in developing and evaluating the tech-nology in relation to current and imagined work activities would be aided by designer participation since designers would bring knowledge of technology constraints and opportunities to the collaboration. The success of the project would be evaluated on the basis of how well the technology supported the work activities.[8] In this last situation the ethnographer would adopt, in part, the designer's orientation of seeking to understand human behavior insofar at it enabled the design of artifacts better suited to the needs of the users.

TRADITIONAL APPROACHES

To put the preceding discussion of ethnographic field methods in perspective, it is useful to describe some traditional approaches used to provide designers with some understanding of user needs and behaviors. Following these descriptions we discuss differences between these approaches and ethnography.

---

[8]Wolcott (1990) questions whether such an undertaking should be considered ethnography at all since, ". . . the research process deserves the label ethnography only when the intended product is ethnography (e.g., some written account or cultural interpretation)" (p. 47). Van Mannen (1988) makes a similar point in distinguishing between doing ethnographic field work and producing an ethnography: "Ethnography as a written product, then has a degree of independence (how culture is portrayed) from the field work on which it is based (how culture is known)" (p. 4). In situations where there never was the intention of developing a written account of the practices studied other than as required by the design effort, can we truly consider such activity ethnography? If not, how is the field work that accompanies such a technology development effort different from more traditional ethnographic studies?

Customer Surveys

Customer surveys involve administering a standard questionnaire to customers about potential, or existing, products with the intention of determining customer preferences for certain technologies and technology features. These surveys attempt to provide information about the nature of the customers' work in relation to the technology in question. For example, customers may be questioned about such things as the technologies currently in use, the frequency with which they use them, the type of tasks they perform on them, and how they view new features or technologies. Customer surveys typically are conducted by market researchers either at the very early or late stages of product development. The stated advantage of customer surveys is that they provide responses from a large, representative sample of customers in a short amount of time. Typically, the design community receives a tally of responses and a summary statement outlining the results of the survey.

Operability Assessments

This approach involves asking potential users of a product to perform several tasks or operations using a simulator or working prototype. Users are asked to perform tasks designed to test features of the user interface. Measurements are taken to record the amount of time required to complete a task, the frequency with which tasks were successfully completed, etc. Problem areas are identified and highlighted as the test proceeds. Designers not involved in the operability test are given user performance statistics, short descriptions of problems encountered, and potential solutions to identified problems. Operability assessments are usually conducted in a laboratory setting by members of the design community, most frequently at very late stages of product development.[9]

Focus Groups

Focus groups bring together individuals from a cross-section of the customer base to evaluate products and product concepts in a discussion group format. Market researchers work with designers to obtain a description or characterization of the product or concept under review. While designers may provide input on topics to be covered in the focus group, they rarely participate directly. Focus group discussions usually include some consideration of the participants' work activities, environment, and future needs. Although the unedited focus group videotapes sometimes can be obtained for independent analysis, most frequently

---

[9]Blomberg (1988b) notes, "While [operability tests] may reveal some problems with the user interface design or with the functionality provided by the technology, they fail to recognize that the laboratory environment has distinct characteristics which differ in significant ways from the environments in which the technology will be used on a day-to-day basis. . . . . in the laboratory setting the user has none of the social resources that are available in most work environments."

designers are given videotaped summary documents which highlight significant issues talked about during the focus groups.

Designers and market researchers recently have been exploring the use of storyboards, scenarios of use, mock-ups, models, and concept videotapes to elicit comments from focus group participants about the relevance of new technology concepts to their work. Customers are asked to react to these representations of technology concepts from the standpoint of the ways in which these technologies might support current work or might enable them to accomplish work in the future. The use of such representations addresses the problem of how to provide adequate verbal descriptions of a technology concept such that focus group participants might imagine uses for the technology.

Field Trips

Field trips provide designers with one of their only opportunities to observe and interact directly with users in the users' workplace. Field trips typically are of two kinds; field visits and field tests.

1. Field Visits.   Field visits are intended to familiarize product developers with customers and the use of products within particular market segments. These visits, typically lasting no longer than two hours at each of several customer sites, take place anytime during the design cycle. The product development teams help set up the visits but, once on site, designers and other team members interact on their own with customers. Questionnaires, interviews, and brief observations are used to obtain information about the use of current products at the site, as well as possible future customer requirements.

2. Field Tests.   Field tests usually take place following the placement of a new product in the customer's work place and are used to gauge the success of a product as well as to identify opportunities for its improvement. Such tests are conducted by the organizations responsible for product sales and installation, with support from the product development teams. Questionnaires and informal interviews are used to obtain users' views on product performance. The results of the tests are made available to the product development teams.

Each of the traditional approaches sketched above has strengths and weaknesses. Some are more appropriate in the early phases of the design cycle, when knowledge about the customer and potential uses of a product can be used to shape the characteristics and definition of the product. Often in later stages of the cycle, the designer can do little more than verify that the design is acceptable to users or make minor changes to those aspects of the product with which the users are most uncomfortable. Issues of cost, impending design schedule, and the availability of key people all are important factors that can influence the selection of one approach over another.

Contrasts Between Traditional and Ethnographic Approaches

Traditionally, the user interface designer's role has been limited to making the functionality of the technology available to the end-user. Ideally, designers might be considered user-advocates in that they represent users as the human element in the system by articulating their needs and requirements in negotiations about design solutions with the development team. All too often, however, the only contact designers have with users is in laboratory operability tests or after a product is placed in the user's environment. Since the user is out of the loop during critical phases of design, the designer must represent the user based on experience and imagination, creating a sort of "virtual" user.

By contrast ethnography provides for an ongoing relationship with users based on designers' firsthand knowledge of the users' work setting. The following are some of the ways traditional approaches to understanding users and ethnography differ.

The Context of the Designer-User Interaction

All of the approaches listed earlier, except field trips, are conducted *outside* the users' work place. Product concepts, prototypes or early engineering models are viewed and evaluated in an imagined setting, often utilizing hypothetical tasks. The quality of the information is limited by the designer's ability to represent the concept in relation to some imagined workplace and the users' ability to envision themselves using the product.

The Focus is on the Technology, not the Work

Traditional approaches to understanding user needs are largely technology driven. The focus is on obtaining answers to specific questions about the acceptability of a particular technology concept, prototype or product, rather than on understanding the relationship between the technology and the work it is designed to support. Technology-focused techniques provide little opportunity for designers to learn about the everyday work practices of potential users.

Users are not Collaborators in the Technology Development

Traditional approaches provide little room for collaboration between designers and users over the evolving design, but instead rely on the users' ability to verbalize their needs or to expose inadequacies of the design in isolated *tests* which take place on single occasions (e.g., in focus group discussion or operability tests).

Ethnography on the other hand has the potential of providing a context wherein mutual understanding between users and designers can evolve. Armed with knowledge of user work practices gained through direct observation of users at work, designers are in a much better position to accurately, and more fully, incorporate users' perspectives in the design, with the potential of improving existing products as well as identifying opportunities for new products. However, incorporating the ethnographic approach into design and product development efforts requires some reorientation. Designers must develop skills in interviewing, observation, analysis, and interpretation, while development teams must be willing to shift their emphasis to support early and continued user involvement. These are considerable investments, but the potential benefits also are high.

A PROJECT TO LINK ETHNOGRAPHY AND DESIGN

Researchers at the Palo Alto Research Center (PARC) and designers in the Industrial Design/Human Interface (ID/HI) department at Xerox recently began collaborating to explore new ways of directly linking ethnography and design. This project, the Participatory Design (PD) Project, brings together individuals with backgrounds in anthropology, graphic design, human factors, and industrial design. Together with users from selected field sites, they form the Participatory Design team.

Goals of the Project

The PD project has multiple, interrelated goals which involve understanding user work practices, developing new ways of incorporating such an understanding into everyday design practice, and integrating the lessons learned from this project into Xerox product development.

Characterizing User Work Practices

The PD project is concerned with giving designers new and better ways of gaining an understanding of users' everyday work practices, with a focus on the relation between technology and human activity. Of particular concern are the ways current technologies support work activities and how work practices integrate a collection of technologies into a system of activity. The application of ethnographic field methods is the vehicle by which such an understanding is being developed.

Developing New Design Practices

The PD project also is aimed at changing current design practices to allow for user participation from the beginning of product design. The use of ethnographic-style field work places designers in the users' work setting which provides an opportunity for continued involvement of users in the design process and for design iteration in relation to actual situations of use.

Integration With Product Development

The PD project also is focused on building on the experiences of this project to help shape new, more participatory product development processes throughout the company. The strategy is to introduce the benefits of a participatory approach to product design by example and by involving individuals from other organizations (Marketing and Product Planning) in some of the work.

New Ways of Working

The PD project has required that designers learn new skills for acquiring an understanding of users' work practices and that anthropologists learn new, non-text based, ways of representing the insights gained from an ethnographic study. Because most traditional approaches to understanding users are not field work based, designers involved in the project wanted some grounding in ethnographic field methods before they embarked on this project. Two workshops were held on the topics of Ethnographic Field Methods and Qualitative Data Analysis. The first workshop provided designers with a perspective on ethnography, as well as practical skills related to observing work practices, conducting open-ended interviews, and making audio and video recordings in the field (see Appendix 1 for field exercises used in this workshop). The second workshop focused on ways of analyzing and interpreting the information acquired from interviews, observations, and video recordings.

Scope of the Project

The PD team has worked primarily with one user community that was selected because (a) a broad range of technologies (fax, computers, printers, typewriters, etc.) were in use at the site, (b) the work was shaped by a rich array of documents (both paper and electronic), (c) groups were linked through a variety of media, and (d) people at the site were eager to take part in the project. This combination of characteristics enabled the team to explore how information was organized and disseminated, how various activities were coordinated, how documents helped structure activities, and how work practices integrated stand-alone office technologies into a coherent system. These were all important issues for the kinds of technologies members of the PD team were being asked to design.

VIDEO ANALYSIS "WALL"

**Layered documentation of ideas, issues
and opportunities for change resulting
from video analysis**

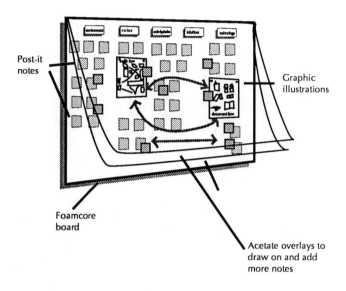

FIG. 7.4.   One of the techniques developed by designers to aid in video analysis.

The team began its field work by interviewing users. These open-ended interviews served as introductions to people, activities, and technologies. During these interviews users were asked to give a guided tour or walk through of their offices, which included a description of the artifacts they used, the people with whom they interacted, and the activities in which they engaged. The interviews were videotaped and were the basis upon which the team chose one group to observe more closely and involve in the actual codesign effort.

THE "ENVISIONER"

A 3 dimensional representation of an office layout

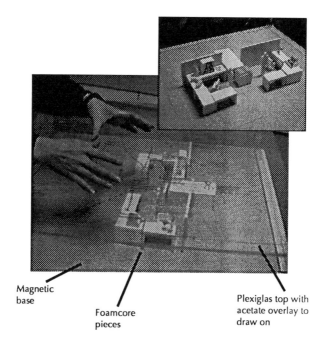

Magnetic base

Foamcore pieces

Plexiglas top with acetate overlay to draw on

FIG. 7.5.   Photograph of industrial design techniques adapted to work practice analysis.

The activities of the group selected were observed and videotaped over several weeks. By jointly analyzing both the interview and observational videotapes, the PD team began to build a shared understanding of the work at these sites. Drawing on skills in graphic and industrial design, the team began to construct a collage of ideas, issues, and opportunities for change that surfaced during the joint video analysis (see Fig. 7.4). The collage helped the designers articulate and translate what they were learning into possible design concepts.

At the same time the team was developing new ways of representing both the work at the sites and possible new technologies to support it. They explored the use of concept videotapes, 3-D representations, graphical depictions, storyboards, and scenarios of use to represent possible relationships between work and emerging technology concepts. These representations aided communication among team members and between them and users. One such representation tool developed by the team was the Envisioner, which allowed a 3-D scaled (¼ scale and ½ scale to show different levels of detail) model of the studied work settings to be constructed (see Fig. 7.5). The layout of particular offices and the equipment and other artifacts used were depicted with foam core pieces which had magnetic bases that sat on a magnetic grid and could be easily moved and rearranged. A sheet of acetate sat on top of a Plexiglass roof so designers could make annotations, draw connections between objects, people, technologies, and explore design ideas. The Envisioner supported looking beyond the design of a particular technology by making the relationship between work, people and technology the focus of the representation.

The Participatory Design project continues at the time of this writing with user co-design sessions being the current focus of activity. The team has agreed on a technology design direction and is now exploring possible design solutions with users. Concurrent with this activity members of the team are becoming involved as work practice analysts and participatory design experts in some mainstream Xerox product development projects. With little time to reflect on their recent experiences, project members are being asked to adapt their new participatory design skills and practices to the requirements of product development. While the value of closer, more extended contact with users is beginning to be understood outside the small participatory design project described here, how successful we will be in migrating the approach into Xerox product development projects remains to be seen.

CONCLUSION

Linking ethnographic field methods and design has the potential both to provide designers with new ways of gaining a deeper understanding of user work practices and to provide a context for designers to collaborate with users over the design of new technologies. However, realizing the benefits of a link between ethnography and design presents many challenges including learning how to translate the insights from an ethnographic study into terms relevant to design, providing designers with the skills necessary to be reasonably accomplished field workers, and altering the mind set of product planners and developers so that extensive, in-depth user involvement is viewed as necessary throughout the design and development process. It is our hope that this paper, and the PD project reported upon here, will help move us closer to a successful linkage between the two undertakings.

REFERENCES

Agar, M. (1980). *The professional stranger.* New York: Academic Press.
Agar, M. (1986). Speaking of ethnography. *Qualitative Research Methods, 2,* 11–78.
Arensberg, C., & Niehoff, A. (1971). *Introducing social change: A manual for community develop-ment.* Chicago: Aldine/Atherton.
Atkinson, J. M., & Heritage, J. (Eds.). (1984). *Structures of social action: Studies in conversation analysis.* Cambridge, UK: Cambridge University Press.
Blomberg, J. (1987). Social interaction and office communication: Effects on user evaluation of new technologies: In R. Kraut (Ed.), *Technology and the transformation of white collar work* (pp. 195–210). Hilldale, NJ: Lawrence Erlbaum Associates.
Blomberg, J. (1988a). The variable impact of computer technologies on the organization of work activities. In I. Greif (Ed.), *Computer-supported cooperative work: A book of readings* (pp. 771–781). San Mateo, CA: Morgan Kaufmann.
Blomberg, J. (1988b, February). *Social aspects of operability: Ethnography of photocopiers.* Paper Presented at the American Association for the Advancement of Science Annual Meeting, Boston.
Blomberg. J. (1989). *Workshop on ethnographic field methods and their relation to design.* Present-ed to Xerox Industrial Design/Human Interface Department. Rochester, NY.
Blomberg, J., & Henderson, A. (1990). Reflections on participatory design: Lessons from the Trillium community. *In* J. C. Chew & J. Whiteside (Eds.), *Human factors in computing systems,* CHI '90 Conference Proceedings (pp. 353–359). New York: ACM.
Briggs, C. (1986). *Learning how to ask: A sociolinguistic appraisal of the role of the interview in social science research.* Cambridge, UK: Cambridge University Press.
Cicourel, A. (1982). Interviews, surveys, and the problem of ecological validity. *American So-ciologist, 17,* 11–20.
Clifford, J. (1988). *The predicament of culture: Twentieth-century ethnography, literature, and art.* Cambridge, MA: Harvard University Press.
Clifford, J., & Marcus, G.E. (Eds.). (1986). *Writing culture.* Berkeley: University of California Press.
Council of the American Anthropological Association (1971). Statements on ethics: Principles of professional responsibility. In J. Cassell & S. Jacobs (Eds.), *Handbook on ethical issues in Anthropology* (pp. 96–100). Washington, DC: American Anthropological Association.
Crane, J., & Angrosino, G. (1974). *Field projects in anthropology: A student handbook.* 2nd edition. Prospect Heights, IL: Waveland Press.
Ehn, P. (1988). *Work-oriented design of computer artifacts.* Stockholm: Gummessons.
Geertz, C. (1973). *The interpretation of cultures.* New York: Basic Books.
Geertz, C. (1983). *Local knowledge: Further essays in interpretive anthropology.* New York: Basic Books.
Goodwin, C., & Heritage, J. (1990). Conversation analysis. *Annual Review of Anthropology, 19,* 283–307.
Greif, I. (Ed.). (1988). *Computer-supported cooperative work: A book of readings.* San Mateo, CA: Morgan Kaufmann.
Harris, M. (1979). *Cultural materialism: The struggle for a science of culture.* New York: Random House.
Heritage, J., & Atkinson, J. M. (1984). Introduction: In J. M. Atkinson & J. Heritage (Eds.),

Structures of social action: Studies in conversation analysis (pp. 1–15). Cambridge, UK: Cambridge University Press.

Jackson, J. (1990). "Deja entendu": The liminal qualities of anthropological fieldnotes. *Journal of Contemporary Ethnography, 19,* 8–43.

Jordan, B. (1987). *Method in sociocultural anthropology.* Course notes, Michigan State University.

Jordan, B., Henderson, A., & Tatar, D. (in preparation). *Interaction analysis: Foundations and practice.* Palo Alto: Institute for Research on Learning and Xerox Palo Alto Research Center.

Kluckhohn, C. (1940). *The conceptual structure in middle American studies. The Maya and their neighbors.* New York: Appleton-Century.

Mehan, H. (1979). *Learning lessons: Social organization in the classroom.* Cambridge, UK: Cambridge University Press.

Mishler, E. G. (1986). *Research interviewing: Context and narrative.* Cambridge, MA: Harvard University Press.

Nader, L. (1974). Up the anthropologist: Perspectives gained by studying up: In D. Hymes (Ed.), *Reinventing anthropology* (pp. 284–311). New York: Vintage Books.

Naroll, R., & Cohen, R. (Eds.). (1970). *Handbook of method in cultural anthropology.* New York: Columbia University Press.

Niehoff, A. (1966). *A casebook of social change.* Chicago: Aldine.

Pelto, P. J. (1970). *Anthropological research.* New York: Harper and Row.

Penniman, T. K. (1974). *A hundred years of anthropology.* New York: Morrow.

Strathern, M. (1987). The limits of auto-anthropology: In A. Jackson (Ed.), *Anthropology at home* (pp. 16–37). London: Tavistock Publishers.

Suchman, L., & Jordan, B. (1990). *Validity and the collaborative construction of meaning in face-to-face survey interviews: An interaction analysis.* Palo Alto: Institute for Research on Learning Report No. IRL 90-0019.

Suchman, L., & Trigg, R. (1990). Understanding practice: Video as a medium for reflection and design. In J. Greenbaum & M. Kyng (Eds.), *Design at work: Approaches to collaborative design* (pp. 65–89). Hillsdale, NJ: Lawrence Earlbaum Associates.

Tyler, S. (1979). *The said and the unsaid: Mind, meaning and culture.* New York: Academic Press.

Van Maanen, J. (1988). *Tales of the field.* Chicago: University of Chicago Press.

Whiting, B., & Whiting, J. (1970). Methods for observing and recording behavior: In R. Naroll & R. Cohen (Eds.), *Handbook of method in cultural anthropology* (pp. 282–315). New York: Columbia University Press.

Wolcott, H. F. (1990). Making a study "more ethnographic." *Journal of Contemporary Ethnography, 19,* 44–72.

APPENDIX 1: FIELD EXERCISES

The following is a list of field exercises developed for a workshop on Ethnographic Field Methods. Although these exercises are most profitably employed in conjunction with a general discussion of ethnography and ethnographic field methods, they are presented here to provide some ideas for practical activities in which people interested in developing skills in ethnography might engage (from Blomberg, 1989: adapted from Crane & Angrosino (1974) and Jordan (1987).

1. Through observation and careful note taking investigate how space is employed in some area of this building. Explore the extent to which the area you

study and the furniture within it tend to keep people apart or to draw them together. How does the arrangement of walls, doorways, hallways, furniture, machines, etc. effect the patterns of interaction you observe?

2. Construct a map of an office area. Your map should show where people sit, their relationship to one another, the activities that go on at various locations, where "significant" objects are located. Describe how you gathered the information represented on your map (through observation, interview, available documents, etc). You'll probably want to include a key or legend to help others interpret your map.

3. Find a specialist (accountant, machine operator, drafts person, etc.) who would be willing to talk to you about the work they do. Ask them to show you how they accomplish some aspect of their work (complete a transfer of funds, prepare and distribute a memo for a group meeting, create a drawing). On the basis of your observations and questioning of the specialist, prepare a description of how the activity is accomplished, the materials used, the people consulted, etc. so that someone who is completely unfamiliar with the activity would be able to get a good idea of what's involved. The object of this exercise is to sharpen your powers of observation and description, and to develop your ability to ask more and more precise questions of your subject. Audio record the interaction.

4. Collect the work life history of someone working in your organization. Collecting life histories requires that the interviewer be as nondirective as possible, interrupting as little as possible, and allowing the person providing the information to decide what's important to tell. You might want to start by asking a nondirective question like, "Tell me about the jobs you've had?" If the conversation falters, you might ask about specific jobs, what was involved in each, why they moved on, what they learned at each position, what they hope to be doing in the future etc. Audio record the interaction so that you are able maintain an appropriate interactional relationship with the subject.

5. Do a semantic analysis of the kinds of documents found in someone's office, or the kinds of telephone calls they make or receive. Ask them to give you the name of each type of document found in their office or the calls they make or receive. If they don't have a name for some category of documents or calls, just include a description of the category. Be careful not to impose you're own categories and names. Ask if there are sub categories within particular classes (personal call of less than 2 minutes, routine requests, etc.) Ask them the characteristics of each document or call included in a particular category and ask them to explain how these characteristics are used to distinguish one type of document from another. If possible, locate actual documents or inquire about actual calls made or received while you were present and ask which category each falls into and why. Their idealized classification system may need to be adjusted to accommodate actual instances. If you have time you might want to

repeat the procedure with another informant and then compare their classification systems.

6. This exercise should be coordinated with exercise 7. Locate a public area where a video camera can be set up. Decide on the best location for the camera. Make note of the issues you considered when deciding where to locate the camera (facial expressions captured, lighting, movement in and out of the space, close-up shots of buttons pushed etc.). Record no more than 30 minutes of the activity at the chosen location. While the camera is running, observe what's going and take notes, paying particular attention to things the camera is unable to record.

7. In the same location selected for exercise 6, take 10 photographs. Choose the shots so that they are representative of the activities and the physical arrangement of objects relative to one another. Attempt to capture interaction and the physical and social context of the activity. Describe each photograph (when it was taken, what it represents, its relationship to the other photographs, etc.). Together with the person doing exercise 6, compare what you were able to capture with the video camera, your observations, and the still photos.

8. Locate an area where you can unobtrusively observe some activity. Observe the activity for 30 minutes or so. Take detailed notes of the what's going on, including a sketch of the area, time of day, movement in and out of the scene, participants' relation to one another. You might want to develop a table to facilitate recording your observations. Review your notes and prepare a description of what you observed.

9. Interview someone with the goal of developing a description of their "social network." Describe the working relationship (lines of authority, relationship between tasks each engages in, etc.) that exists among co-workers. Represent this information in an "organization chart" where the nature of the relationship between co-workers is represented.

10. Go to a public area where people meet (cafeteria, coffee room, etc.). Select a small group of people to observe (2–6). Based only on your powers of observation try and determine who these people are (age, occupation, education, etc.), their relationship to one another, what they're talking about, etc. After observing them for a few minutes and writing down your hunches, interview them and find out to what extent your impressions were related to what you learned from interviewing. Reflect on how you misread certain cues, the information unobtainable through observation alone, etc.

11. Interview someone about the "life history" of some small collection of documents (i.e., the ones currently on their desk or in their "in" basket). Talk with other people involved in creating, processing, or acting upon the documents. Prepare a description of each document's life history.

8

Cooperative Design: Techniques and Experiences From the Scandinavian Scene

Susanne Bødker
Kaj Grønbæk
Morten Kyng
Aarhus University, Denmark

This chapter is based on a vision, an ideal, of what system development should achieve and how it should take place. First of all, we see the ideal project as one that encourages the users-to-be in an organization, i.e., all the involved groups and individuals, to decide themselves how to develop their work by means of new computer support. This ideal is seldom realized, because organizations are characterized by conflicts, most significantly between management and labor, and because different groups have different power and resources in the organization. It is not only a matter of democracy, though. In our experience, most traditional systems development methods are too abstract and detached from the work and life of most people, even from management. Thus we need to develop and work with techniques that allow for more use of everyday experiences and professional skills in design.

At the Participatory Design Conference (PDC '90), Seattle, the authors of this chapter and their collaborators were invited to present their approaches to participatory design, together with other researchers and designers from Scandinavia. These approaches originate from a diversity of projects where researchers and designers have been cooperating closely with groups of workers, shop stewards, and other union representatives, and where the organizational framework provided by the local unions played an important role (Ehn & Kyng, 1987). Hence, one of the key questions discussed over the 2-day conference was whether the Scandinavian "Model" for system design, as summarized in our earlier vision could be applied in a North American setting—a setting where industrial democracy and strong union relationships cannot be assumed to be found in most organizations, as is the case in Scandinavia. We do not attempt here to answer this question. Rather, we do discuss a set of techniques to support our earlier vision. We also describe the underlying perspectives and leave it to the readers to judge whether the perspective and the techniques can be adopted in their particular settings. For detailed discussions of the techniques, and the rationale behind them, we refer to a recent book, *Design at Work: Cooperative Design of Computer Systems* (Greenbaum & Kyng, 1991).

To begin, we present our perspective on computer applications and design as a list of statements or assumptions:

Computer Applications

• When computer applications are brought into a workplace, they should enhance workplace skills rather than degrade them.

• Computer applications should be viewed as tools, and designed to be under the control of the people using them. They should support work activities, not make them more rigid.

• The introduction of computer applications changes the organization of work around them. The interplay between the computer application and work organizational issues should be a specific focus of the design and introduction of computer applications into organizations.

• Although computer applications are generally ordered to increase productivity, they also need to be looked at as a means to increase the quality of the results.

The Design Process

• The design process, as any process taking place in an organization, is a political one and leads to conflict. Managers who order an application see things differently from the workers who will use it. Different groups of users will need different things from the application, and system designers often pursue their own interests. Conflicts are inherent in the process. If they are ignored the solution may be less useful and continue to create problems.

• Computer applications that are created for the workplace need to be designed with full participation from the users—both from a democratic point-of-view and to insure that competencies central to the design are represented in the design group. Full participation, of course, requires training and active cooperation, not just token representation in meetings or on committees. We use the term *cooperative design* to designate such cooperation between users and designers. However, to users, designing a new computer application is a secondary activity whereas for designers it is their primary work. This means that the designers should know how to set up the process and need to make sure that everyone gets something out of the interaction.

- The design process highlights the issue of how computers are used in the context of work organization. We see this question of focusing on how computers are used, which we call the *use situation,* as a fundamental focus for the design process. We put our attention on how people work, and take the practice of the users as the starting point for the design process.

- Encouraging user participation and designing for skill means paying attention to things that are often left out of the formal specifications, like tacit knowledge or shared knowledge and communication. Computer applications are a lot more than the simple flow of information represented in the flowcharts that systems analysts present to their clients. When users participate in actual design activities it is necessary to use tools that are familiar to them. Traditional tools such as flowcharts, dataflow diagrams, and programming languages are insufficient (or even useless) as means for cooperating with users. In our experience, some of the techniques we discuss later do in fact encourage user participation and creativity.

- To enable users to contribute with their tacit knowledge in design, it is important to simulate future work situations, creating the illusion of actually working with the projected system. In this way changes in the use practice can, to some extent, be predicted and evaluated.

This perspective on computer applications and design processes has evolved over more than 15 years, strongly influenced by the experiences from several research projects of our own as well as those of other groups. An overview of this development follows.

HISTORICAL BACKGROUND

The research projects that we have been involved with since the early 1970s, have developed in interplay with the society around them. Employee influence through unions and cooperation with management has for a long time been an integral part of the industrial relations in the Scandinavian countries. The Scandinavian projects on worker influence on development and use of computers began in the early 1970s, when new legislation increased the possibilities for worker influence. The projects supplemented the earlier strategy, called codetermination, with a series of activities set up by central and local unions independent of employer organizations and management. In these "first generation" projects, workers aided by consultants and researchers struggled to develop a better, more coherent platform for worker influence on the use of new technology at the workplace. New work practices, focusing on group work and the development of local resources for action, were being shaped, tried out in practice, and reshaped in the projects. Some of the work groups produced criteria for better working environments and suggestions for applications to support groups of workers planning their own work. As a result of the first of these projects the existing legislation on worker influence was supplemented by Technology Agreements that gave workers a direct say in the development and use of technology in their workplaces. This also led to an extensive series of union education programs. The first of these projects was the NJMF project, set up by Kristen Nygaard and Olav-Terje Bergo together with the Norwegian Iron and Metal Workers' Union (Nygaard & Bergo, 1975). This was followed by the Swedish DEMOS project in which Pelle Ehn and Åke Sandberg played major parts, and the Danish DUE project organized by Morten Kyng, Lars Mathiassen, and Niels Erik Andersen together with the Trade Union Council (Ehn & Kyng, 1987).

In the 1970s these early projects introduced the notion of worker participation in decisions about technology, but, whereas workers had a legal say in workplace technology, the laws did little to shift the balance of power from a managerial perspective. And the rationalistic tradition embedded in computer system development did little to give workers a voice in putting forth their own ideas when trying to agree on the introduction of new technology. As in the United States, this was reflected in the tools of system development, which emphasized developing technical specifications rather than seeing the application from the perspective of the users.

By the early 1980s, a "second generation" of projects was initiated in Scandinavia. These projects focused on design of new kinds of computer support using skill and product quality to push computer system design more towards a users' perspective. They took the issue of dehumanization and put it on the table as a central problem in the design and use of computer systems. Thus, to put some muscle on the bones of the Technology Agreements, the issues of quality of work and product were put into the foreground of the new design-oriented projects. An example of this was the Utopia project, named both for its ideals and as an acronym for its use (Bødker, Ehn, Kammersgaard, Kyng, & Sundblad, 1987). In this project computer system developers and researchers worked with a group of typographers helping to formulate the ways that computer technology could be used to enhance their skills and better the typographic quality of newspapers. These researchers came from the Swedish Center for Working Life, the Technical University in Stockholm, and from Aarhus University.

From the first projects in the 1970s, developing strategies and techniques for worker influence on design and use of computer applications, through the focus on skill in the early 1980s, computer system development in Scandinavia developed the concept of user participation. In *Work Oriented Design of Computer Artifacts,* Pelle Ehn (1989) outlines the story of these changes and delves into some of the theoretical work that influenced the thinking of the earlier as well as later projects. Rethinking the actions and experiences in these projects he develops a theoretical understanding of design work based on phenomenology, marxism, and ordinary language understanding.

This theoretical understanding has been supplemented with inspiration from activity theory (Bødker, 1991; Christiansen, 1988), and the so-called work development research (Engeström, 1987; Bisgaard et al., 1989). Together with the insight developed during our writing of *Design at Work* (Greenbaum & Kyng, 1991) it has lead to a new round of empirical work, which we present in part next.

In retrospect, when looking at the strategies and techniques applied in the aforementioned projects, there has been continuous development and some major changes: In general, we have today a much better theoretical understanding of possibilities for and limitations to active user participation in design processes. This understanding comes from our interpretations of theories that can be grouped under the philosophical heading of social construction, seen in contrast with the rationalistic tradition of computer science. At the practical level this is reflected in our continued *revisions* of our set of development tools and techniques, where we try to modify those aspects of our techniques that are influenced by the older rationalistic perspective pervading system development theory and practice. One illustration of these revisions is our move from system descriptions to mock-ups and game-like design sessions (see the following). Also, where many of the techniques of the early projects followed the line of analyzing, followed by setting up goals, for example, for local unions, and then making action programs, today, many of our techniques use fantasy and imagined futures to move from problems in the present specific setting to reality and specific actions. This "substitution" of goals with positive visions/fantasy without the need for rationalistic arguments to support the visions is an important step in encouraging creative contributions from the users (Kyng, 1989).

CASE: THE AT PROJECT

The following case example is a project taking place at the National Labour inspection service (AT) office in Aarhus. The purpose of the project seen from the point-of-view of the managers and workers at AT is to design a number of computer applications for AT and to develop a long-term strategy for decentralized development and maintenance. Decentralization here is relative to the existing strategy in which all AT offices in Denmark subscribed to centralized computer applications run by a department in Copenhagen. This aspect, that the project is useful for the people in the participating organization, is central to our research. As with all the empirical projects that researchers from our tradition engage in, the project is one of action research, which in our understanding, means that the people we work with should get something out of the process as well. The purpose for the researchers is an improved understanding and further development of the techniques described in *Design at Work,* among others, and of the applied theoretical basis.

We use the case primarily for presenting our choice of techniques, based on experiences from previous cases. The AT project itself is still only in the first stages. Presenting the techniques by means of a concrete case also serves to illustrate the situated nature of our techniques. The way we combine and apply the techniques at AT is dependent on specific conditions at AT. So, for instance, the techniques are presented differently in *Design at Work.* And obviously, the way we planned the work at AT differs from what actually happened.

In the project, 8 researchers work together with the people at the AT office in Aarhus. This office is part of a country-wide labor Inspection service, under the auspices of the Danish State, mandated to take care of the workers' health inspection at all workplaces. Employed are inspectors, who are partly specialized in different areas, as well as administrative personnel, and a centrally located support staff of researchers. In the Aarhus office, the staff amounts to approximately 40 people.

The inspectors are trained in many different ways, as engineers, craftspeople, biologists, nurses, medical doctors, psychologists, and more. The contact between the inspectors and the companies takes place in three different ways: (a) an inspector selects, and visits, a randomly chosen company; (b) an accident is reported by a company, a doctor, or a shop steward, upon which the inspector may choose to pay a visit; (c) and through a so-called campaign. A campaign is a coordinated (often from the main office) effort to visit companies within a certain area, or where certain kinds of work hazards are very common. For example, all plastic-producing factories are paid visits to inform about cancer. The inspectors in general give information and advice, and they have the legal power to order changes in the production facilities when these are not in accordance with the law.

The administrative staff consists of office workers, who perform numerous support functions, maintain files of different kinds, and give immediate help to people calling on the phone.

For the researchers, this project is somewhat different from most previous projects in that we cooperate both with the local management and with the workers. Furthermore, the project from the outset includes all groups at the organization, with their different roles and skills.

Following are presentations of the techniques that the project works with as outlined in Fig. 8.1. We also discuss specific measures. The following activities were planned during the first part of the AT project:

• Researchers learning about the work at AT through workplace visits during which people from AT were interviewed and demonstrated their work practices.

• Compilation of an inventory of existing problems with and new ideas for

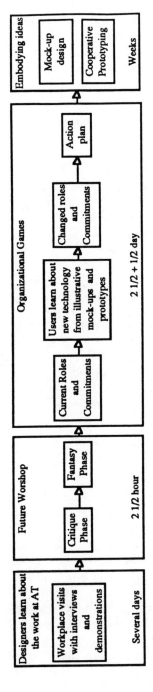

FIG. 8.1. Activities undertaken and planned for the AT project.

work organization and computer support through the critique and fantasy phases of a Future Workshop.

- An Organizational Game investigating current roles, new technological possibilities—illustrated by means of mock-ups and prototypes—and changed roles, and developing an action plan.
- Embodying ideas—continued cooperative design of selected computer applications using mock-up design and cooperative prototyping, and trying out new/modified work organizations.

At this point, a detailed plan for development of a strategy for decentralized system development or for the conclusion of the project has not been made.

The initial step of the first activity was for 4–5 researchers to visit the organization for a couple of days. We wanted to learn more about the work tasks of the inspectors and the administrative staff. Furthermore, we needed to focus on the materials and instruments that were applied in the different work tasks, and the cooperation between people in the work tasks. This involved following several of the inspectors on workplace inspections. In general the strategy was one of interviewing people in their offices or wherever they were working, asking them to demonstrate to us parts of their work tasks, the files that they keep, the instruments, in particular computer programs that they apply, etc. Inspired by Engeström (1987) and Bisgaard et al. (1989) focus was not only on the normal state of affairs, but also on exceptions and problems. The work of managers and workers was treated in the same way, and the outcome was improved knowledge of the organization as well as a number of ideas and problem areas to focus on in the process to come.

FUTURE WORKSHOPS

Following the workplace visits at AT, a Future Workshop was conducted to help focus more directly on the problems and ideas that the workers at AT had, or could bring forward collectively. In *Design at Work* (Greenbaum & Kyng, 1991) the use of the Future Workshop technique for system design is described in detail. (See Halskov Madsen & Kensing (1991) who have developed the Future Workshop technique for system development in combination with a technique denoted Metaphorical Design.)

Robert Jungk and Norbert Müllert (1987) originally developed the Future Workshop technique for citizen groups with limited resources who wanted a say in the decision-making processes of public planning authorities (town planning, environmental projection, energy crisis, etc.). The technique is meant to shed light on a common problematic situation, to generate visions about the future, and to discuss how these visions can be realized. Those participating should share the same problematic situation, they should share a desire to change the situation according to their visions, and they should share a set of means for that change. Usually, a Future Workshop is run by at least two facilitators, with no more than 20 participants. The facilitators attempt to ensure an equal distribution of speaking time and they should also ensure that all participants can follow the discussion, by letting the participants write their ideas as short statements on wall charts (a large sheet of paper taped to the wall). A Future Workshop is divided into three phases: the Critique, the Fantasy, and the Implementation phase. Essentially the Critique phase is designed to draw out specific issues about current work practice; the Fantasy phase allows participants the freedom to imagine "what if" the workplace could be different; and the Implementation phase focuses on resources needed to make realistic changes. These phases are surrounded by preparation and follow-up periods.

At the beginning of the Future Workshop the facilitators introduce the technique to the participants. Basically the Critique phase is like a structured brainstorming that focuses on current problems at work. Speaking time is allotted to all participants and is often restricted to 30 seconds to make it possible for everyone to speak. As a preparation for the next phase the statements are grouped under a number of critique-headings/themes. As the first part of the following Fantasy phase the critique themes are inverted to positive guiding themes. During the following brainstorming no statement about the future is considered too extreme—if somebody wants it, it's OK. As a preparation for the second part of the Fantasy phase the short statements are regrouped under a number of fantasy-headings/themes. Some of these are then selected for elaboration in the following group work, where they are elaborated into "utopian outlines." The Implementation phase may start by having each group present their version of a utopian outline. The general idea is to make plans for how to start acting in the immediate future of the workshop. In the AT project the Future Workshop idea was adapted as follows.

The researchers' understanding of the problems of the organization was mainly based on interviews with individuals at AT. After that, they encouraged workers to also formulate their own understanding of problems and changes, preferably as a collective. Not that they expected everybody to share needs and problems, but at least to create a mutual understanding of these.

Given previous experiences with Future Workshops where different groups were involved (see for example, *Design at Work,* Chapter 7), the researchers decided to ask management to stay away from the workshop and leave the scene to the workers. One of the problems experienced earlier in having a mixed group of managers and workers do a Future Workshop together is that the workers may be afraid of management retaliation if they express serious critique of current conditions in the organization. Also, in many organizations, the workers are less used to expressing themselves orally than management, resulting in a workshop where management interests set the agenda.

It is quite difficult to get everybody together in an organization like AT, where the inspectors are out a lot, and where many people work part-time. Thus, a 2½ hour slot in a regular monthly staff meeting was set aside for a meeting focusing on the Critique and Fantasy phases of a Future Workshop. In addition to this, a 2-day meeting for a smaller group was scheduled for later. For this meeting, the researchers decided to use the design-by-playing technique described shortly. In order to make the best possible use of the 2½ hours, the Implementation phase was seen as something going on after the meeting (in many ways, the rest of the project can be seen as implementation). To make use of the insight already collected two of the four researchers present in the meeting were to observe the statements in the Critique phase and try to come up with themes for the Fantasy phase. This also reflects an experience of ours saying that formulating themes is often hard and time consuming. The Fantasy phase was conducted in four groups, each of which had a different theme and had a researcher as facilitator. Some of the ideas that came up here were selected for further exploration in the Organizational Game. The themes arising out of the Future Workshop were used in setting up the playground of the Organizational Game, and the discussions as well as the earlier investigations were used in setting up the situations for the game.

ORGANIZATIONAL GAMES

The major problems in AT are centered around the organization of work and how this relates to the use of computer technology: For instance, much of the information that the inspectors report about their inspections is rewritten three times by the inspectors in slightly different forms. Following this, the information is filed by different secretaries into different files. It was suggested by the researchers, and accepted by AT, that the 2-day meeting should focus on such work organization problems. The Organizational Game technique, developed by Sjögren and Ehn (1991) seemed appropriate for this. It supports considering alternative work organizations by playing them out and confronting the different problems they create.

Organizational games build on the idea that:

- they make a difference for the participants
- implementation of the results is likely
- they are fun to participate in.

The first two points concern the political side of participation in design; the users must have a guarantee that their design efforts are taken seriously. The last point concerns the design process. No matter how much influence participation may give, it should transcend the boredom of traditional design meetings to really support design as meaningful and involved action. The design work is treated as a theatrical play. Furthermore the organizational games share with future workshops the idea that the overall structure of the game moves from the present situation, via an imagined future, back to reality. One of the early attempts to develop organizational games came from the UTOPIA project where an Organizational Kit was developed to enrich work with abstract system descriptions. The basic ideas behind the organizational design game in the UTOPIA project were that:

- It should be fast and easy for a group of people to work with;
- it should be cheap and flexible to use, allowing several alternatives to be tested during discussions; and
- it should be based on concepts relevant to the actual type of production and support design discussions of existing and future work and technology.

Based on experiences from UTOPIA and other projects, the technique has been developed and applied in several projects, by Sjögren and Ehn (1991). The main example of the use of the technique is the Desktop Publishing Game. It was developed for and played in a few public administration offices in Sweden, the main case being the Consumers Agency Konsumentverket. The technique borrows its way of thinking from theater. Its dramatic design context is based on six concepts:

The playground is a subjective and negotiated interpretation of the work organization in question. The professional roles are represented by both individual professional ambitions and organizational requirements. The situation cards introduce prototypical examples of breakdown situations. Commitments are made by individual role players as actions related to a situation card. Conditions for these commitments are negotiated, and an action plan for negotiations with the surrounding organization is formulated.

The game designed for AT was based on a group of 10–12 people from different parts of the organization, workers as well as management. The game started with a Prologue, where the rules of the game were introduced, and the roles and competences of the participants chosen. For Act 1, which followed the Prologue, the playground and situation cards were designed by the researchers to focus on the problems that had surfaced during the initial activities in the project, i.e., the workplace visits and the Future Workshop. In later parts of Act 1, the participants were to create and play their own situation cards. This was done to bring the game closer to the actual problematic situations of the organization, as perceived by the participants from AT. A lot of work was put into preparing Act 2, where some amount of not-yet-implemented or well-defined technology applications were to appear on the scene. The researchers decided to instantiate this technology by means of prototypes and mock-ups (see the following) letting participants get a chance to experience these as part of the game. Act 2 focused on possible new pieces of technology to be applied in the organization, and how they would affect the work in the organization. Act 3 focused on changes of roles and new commitments, with new technology. This act applied scenarios rather than situation cards. One scenario was the small autonomous group, and one was focused on process and product quality. Once again the roles and competences of the participants were negotiated. In the final part, a new playground was brought in, and the participants negotiated an action plan, based on the experiences from the game. This action plan focused both on what could be done here and now and by groups or individuals in AT, and on what needed external resources. All of this was scheduled to take place at a 2½-day seminar. However, we did not finish the action plan. This was done 10 days later at a half-day meeting.

The following two sections present some of the techniques we have developed to involve users actively and creatively in the design of computer applications. Their main value is that they allow users to experience future work-like situations where emerging new applications are tried out, evaluated, and changed or discarded—in cooperation between users and (system) designers. This use of the techniques is part of the action plan developed in the Organizational Game described earlier, but thus far the techniques have, in the AT project, only been used in a learning process, where the designers illustrated new technological possibilities by means of mock-ups and prototypes as part of the Organizational Game (cf. the second act described earlier). As Fig. 8.1 shows, mock-up design and cooperative prototyping activities are planned to embody a subset of the ideas that surfaced in the previous activities.

MOCK-UP DESIGN

The idea of mock-ups was developed in the UTOPIA project as a way to encourage active user involvement, as opposed to the use of traditional specification documents. In addition, they actually help users and designers transcend the borders of reality and imagine what is currently impossible, such as a screen of 100" with 1000 pixels per inch. As opposed to descriptions, mock-ups remind the users of familiar work situations. And with mock-ups there are meaningful roles for them to play in the use of these design artifacts.

The idea of mock-ups is presented by Ehn and Kyng (1991). They ask why mock-ups work, despite having low functionality and being only a kind of simulacrum. Some of the answers, they suggest, are that mock-ups:

• encourage "hands-on experience," and thus support user involvement beyond the detached reflection that traditional system descriptions allow for;

- they are understandable, hence there is no confusion between the simulation and the "real thing," and everybody has the competence to modify them;
- they are inexpensive, hence many experiments can be conducted without big investments in equipment, commitments, time, and other resources; and last but not least,
- they are fun to work with.

Some of these characteristics are shared by mock-ups and prototypes, as described in the following section. However, some are not, and in the following we focus on these characteristics, that is, why at times it may be a good idea to design without computers. First, mock-ups are built with inexpensive materials. To buy expensive hardware and build advanced software early in a project may, in most situations, be directly counterproductive, especially given the possibilities of mock-ups. In other situations, however, the investments in hardware and software may not be a problem—PCs may already be massively used in the organization. Still, the use of mock-ups may pay off, because it can help generate new visions and new options for use. Second, the characteristics of these simple tools and materials are familiar to everybody in our culture. With this type of mock-ups nothing mysterious happens inside a 'black box.' If a picture taped to the blackboard drops to the floor everybody knows that this was due to difficulties of taping on a dusty chalkboard, and not part of the design. There is no confusion between the simulation and the 'real thing.' Third, such mock-ups lend themselves to cooperative modifications. The possible "operations" on the material using, for example, pens and scissors, are well known to all, and with simple paper-and cardboard mock-ups people often make modifications jointly or take quick turns changing the mock-up. The physical changes are visible, and, with proper display, visible to all the participants.

However, as with any tool or technique, simple mock-ups have their limitations, too. First, changes to a mock-up may be very time consuming. If, for example, a different way of presenting menus is chosen, changes may have to be done to dozens of drawings, or a whole new set of slides will have to be made. Second, it is hard to illustrate behavioral aspects of future applications as realistically as it can be done with prototypes. Third, although mock-ups allow a design group to experiment without the limitations of current technology, this freedom is only a partial blessing. In the end, good design results from a creative exploitation of the technological possibilities and limitations, not from ignoring them. Thus, as paradoxical as it may sound, the demands for computer knowledge in a design group using mock-ups are very high.

In the AT case, we plan to build mock-ups to explore more details of possible future technology, by managers, workers, and researchers in cooperation. Suggestions for where to use mock-ups and prototypes have arisen from the action program, as described next, but the specific areas in which to proceed remain to be chosen.

COOPERATIVE PROTOTYPING

The way we do prototyping—cooperative prototyping—is different from traditional prototyping in that traditional prototyping approaches mainly take the perspective of the developers: Analyst/Designers conduct investigations in the user organization and develop prototypes on their own. Such prototypes are tested by or demonstrated to users to give the developers feedback on their solution. It has a superficial resemblance to our use of prototypes for illustrating new technological possibilities, but whereas we see this use of prototypes as part of the users' learning, traditional approaches view it as part of the feedback to developers. Traditional approaches put little emphasis on active user involvement in the actual design process. Bødker and Grønbæk (1991) introduce a different approach called Cooperative Prototyping. The approach is an exploratory approach (Floyd, 1984) where prototyping is viewed as a cooperative activity between users and designers, rather than an activity of designers utilizing users' more or less articulated requirements. In cooperative prototyping the aim is to have users achieve a familiarity with the tools that comes close to what was achieved with mock-ups, although this may be infeasible in the immediate future.

The cooperative prototyping approach establishes a design process where both users and designers are participating actively and creatively with their different qualifications. A key point in facilitating such a process is to let the users experience prototypes in a fluent work-like situation. The users' current skills must be confronted with new technological possibilities. This can be done in a simulated future work situation or in a real use situation. When breakdowns occur in the simulated use situation, users and designers analyze the situation and discuss whether the breakdown occurred because of the need for training, a bad or incomplete design solution, or for some other reason. Breakdowns caused by bad or incomplete design solutions should rapidly be turned into improved designs, in order to reestablish the fluent work-like evaluation of the prototype. In many cases users can participate actively in improving the prototypes. Prototype modifications that can be made by direct manipulation are made immediately in the session, whereas modifications requiring a larger programming effort are postponed and made by designers after the session.

Setting up alternative prototypes is useful for users and designers to get their imagination going and thereby for the group to discuss different ways of organizing work. A Future Workshop often displays a diversity of alternative fantasies about how to improve work with computers or by changing work organization. Exploring some of these alternatives through prototyping is a fruitful way to move from fantasy into the actual development process and to focus on improving the users' work through computer support.

At present the prototypes for the AT case are seen primarily as learning vehicles for the workers and managers as part of the Organizational Game. This use of prototypes supports imagination and discussion about the future work with new technology. The Organizational Game becomes more concrete than if technology is just talked about. For instance, prior to the AT project, there had been a lot of talk about the inspectors carrying portable computers on their visits. But hardly any of the inspectors had seen one. Letting them experience the general capabilities of a portable computer in itself grounds the discussions. And of course experiencing prototypes of what the portable may be used for in their specific situation brings the discussion much further and also raises a number of questions related to work organization. For instance, if the inspectors directly enter into the computer the information that they now record on forms, what will happen to the work tasks of the secretaries who are now responsible for this data entry?

As with mock-ups these initial discussions help delimit areas in which to proceed with prototyping. And part of this is to let the users get more actively involved with the existing prototypes. To learn the best ways to set up these sessions, a diversity of questions are considered, such as: What is the purpose of the session? How stable should the prototype be in advance? To what extent should in-session modifications be done? What setting should be chosen? How should the outcome be documented/evaluated? In prototyping sessions, designers often like to demonstrate all the features they believe are wonderful. However, our claim is that demonstrations do not necessarily tell the users anything about how the prototype or the final application fulfills their needs. To fully experience the prototype, the users need to be in control of its use for some period of time, to try it out in work-like situations. If the prototypes are not sufficiently stable to let the users work on their own with it, the designer should be prepared to give first aid for breakdowns caused by the prototype. While designers are, initially, the ones who know how the process should be set up, it is important that the process be adjusted to the needs and wishes of the users.

The product of the prototyping process is not only a computer prototype. Prototyping is a learning process, and much of the new understanding must be spread to workers and managers who do not participate directly in the prototyping. One way is to use the different prototypes in a process in which all involved personnel are guided through a compacted version of the prototyping process (Bisgaard et al., 1989). In general, the prototypes are valuable means in the education of future users, because education can start while the final computer application is being implemented. Similarly, the participants from a prototyping process can often act as teachers. These ideas are being developed in the AT project.

GENERAL CONDITIONS

A design process in which a group of users and designers cooperate takes place in a complex context with different types of societal conditions and contractual relationships among the partners (see Chapter 5 by Grønbæk et al., this volume).

System designers in some settings will have to fight to involve users in the design process. For instance, the application to be developed may be specified in a product contract stating requirements and a fixed price. Or the application is a brand new product to be marketed. Thus it may be difficult to identify a suitable group of potential users.

One of the frequent obstacles to increased user involvement in design projects is resources, or rather, the lack of them. Throughout a design project resources must be secured for user participation if there is to be real cooperation. Users who get no reduction in their daily workload cannot be expected to engage themselves in a design project over long periods of time. But resources are not only time and money; they also include education and assistance from a variety of experts.

Another frequent obstacle to user involvement is the current structure and the traditions in the involved organizations. For instance, marketing groups within product development organizations may restrict user involvement in the development process even when the developers do find it important (see chapters by Grudin and by Grønbæk et al., in this volume). One such reason could be that the product development organization does not want to design a product that is tied too closely to a certain customer or customer type.

In most organizations, some groups have more power and resources than others. Those who have the most power and resources are usually management, not the end-users. To help users get a forum where they can take an active part in design means to set up situations where they can act according to their own interests and rules, and not simply according to those of their managers. The right to make decisions about how to spend resources is also a crucial aspect of the resource question mentioned before. Most projects have to follow the rules of the organization to which they belong. In some cases this implies that a project group consisting of designers and users collectively makes the decisions, as was the case with the UTOPIA project mentioned earlier. In most cases, however, standard managerial procedures will determine the way decisions are made. When a crisis or a conflict arises under these circumstances, decisions often run contrary to the wishes of users, and perhaps contrary to those of the designers as well. The result is usually that the users withdraw from the process, either totally by leaving the design group, or by becoming passive members, losing their initial belief in the possibilities of their influencing the design. To give users a better chance of continued participation, we recommend that from the outset the designers argue for a certain amount of resources to be allocated for user initiatives.

The weakest groups in particular need the strongest support in formulation of their demands and ideas for the future. There can be conflicting interests among the groups, but these conflicts may be turned into resources for the project, if the situations are set up in appropriate ways. This can mean that designers and users should not just establish one project group of all involved parties, but should work with different groups of people at different times (Ehn & Kyng, 1987; Ehn

& Sandberg, 1979; Engeström, 1987). At the same time, the different groups need to be exposed to each other's demands and suggestions, which emphasizes the need for designers to be coordinators of the activities.

CONCLUDING REMARKS

One of the issues confronted when we tear down the barriers of traditional design is that the roles of the system developers shift from that of project managers to project facilitators. Cooperative design, which in our perspective means empowering users to fuller participation and cooperation, changes the rules of the game. Within traditional system development each step, from feasibility study through implementation, is supposed to be controlled by system developers and management through discrete procedures, and marked by clear-cut milestones and exit criteria. In our view, the traditional system approach makes it easier for designers to create the impression that things are under control. But often the supposedly last act, where a system is put into operation, shows that it does not fit the work of the users, and an almost unending epilogue of modifications begins. The cooperative design approach begins by creating an environment in which users and designers can actively consider the future use situation. It is a process where users and designers don't have to wait until the final act to know if the application will fit the practice of the users.

Finally, we recall the situated nature of cooperative design. Each application of the techniques described in this chapter will be different depending on the type of project in which they are applied. In *Design at Work* (Greenbaum & Kyng, 1991) the techniques have been presented using a variety of examples of their use; in this chapter we have done a somewhat similar thing using a single project as our example. A reader who examines both sources will see that the techniques have been applied differently. The point is that the next application of the techniques will again be different. This is why we think example driven presentations, like the one here, are more appropriate than stating general guidelines and methods. Trying to apply techniques that have been described by example will force the reader to consider similarities and differences between the described example and the current situation. This exercise will move the attention towards how to fit the technique to the current setting, rather than just sticking to a guideline that most likely doesn't fit the situation.

ACKNOWLEDGMENTS

Many people have been involved in the work on which this chapter is based. Pelle Ehn, Joan Greenbaum, Kim Halskov Madsen, Finn Kensing, and Dan Sjögren were principal "methodology developers" and fellow writers on the

Design at Work book (Greenbaum & Kyng, 1991). Pelle, Joan, Kim, and Merete Bartholdy were members of the group who did the workshops at the Participatory Design Conference held in Seattle, March/April 1990. The researchers in the project that we use as our case include Ellen Christiansen, Pelle Ehn, Randi Markussen, Preben Mogensen, and Randy Trigg, as well as the authors of this chapter. Ellen, Randi, Preben, and Susanne have done many of the initial investigations in the project, and together with Pelle and Randy they have planned and coordinated the workshop activities we discuss. The chapter would not have been possible without the contributions of the many collaborators in the different empirical cases that constitute our frame of reference, including those of the people at AT-Aarhus. Pelle Ehn commented helpfully on an earlier draft.

REFERENCES

Bisgaard, O., Morgensen, P., Nørby, M., & Thomsen, M. (1989). *Systemudvikling som lære-virksomhed, konflikter som basis for organsationel udvikling* [Systems development as a learning activity, conflicts as the origin of organizational development]. (DAIMI IR-88). Aarhus: Aarhus University.

Bødker, S. (1991). *Through the interface: A human activity approach to user interface design.* Hillsdale, NJ: Lawrence Erlbaum Associates.

Bødker, S., Ehn, P., Kammersgaard, J., Kyng, M., & Sundblad, Y. (1987). A Utopian experience. In G. Bjerknes, P. Ehn, & M. Kyng (Eds.), *Computers and democracy: A Scandinavian challenge* (pp. 251–278). Aldershot, UK: Avebury.

Bødker, S., & Grønbæk, K. (1991). Design in action: From prototyping by demonstration to cooperative prototyping. In J. Greenbaum & M. Kyng (Eds.), *Design at work: Cooperative design of computer systems* (pp. 197–218). Hillsdale, NJ: Lawrence Erlbaum Associates.

Christiansen, E. (1988). *Den realistiske vision: et humanistisk-datalogisk perspektiv på systemudvikling* [The realistic vision]. Unpublished doctoral dissertation, Aalborg: Aalborg University.

Ehn, P. (1989). *Work-oriented design of computer artifacts.* Stockholm: Arbetlivscentrum.

Ehn, P., & Kyng, M. (1987). The collective resource approach to systems design. In G. Bjerknes, P. Ehn, & M. Kyng (Eds.), *Computers and democracy: A Scandinavian challenge* (pp. 251–278). Aldershot, UK: Avebury.

Ehn, P., & Kyng, M. (1991). In J. Greenbaum & M. Kyng (Eds.), *Design at work: Comparative design of computer systems.* Hillsdale, NJ: Lawrence Erlbaum Associates.

Ehn, P., & Sandberg, A. (1979). *Företagsstyrning och löntagarmakt.* Stockholm: Prisma.

Engeström, Y. (1987). *Learning by expanding.* Helsinki: Orienta-Konsultit.

Floyd, C. (1984). A systematic look at prototyping. In R. Budde, K. Kuhlenkamp, L. Mathiassen, & H. Zullighover (Eds.), *Approaches to prototyping* (pp. 1–18). Berlin/Heidelberg: Springer-Verlag.

Greenbaum, J., & Kyng, M. (Eds.). (1991). *Design at work: Cooperative design of computer systems.* Hillsdale, NJ: Lawrence Erlbaum Associates.

Halskov Madsen, K., & Kensing, F. (1991). Generating visions: Future workshops and metaphorical design. In J. Greenbaum & M. Kyng (Eds.), *Design at work: Cooperative design of computer systems* (pp. 155–168). Hillsdale, NJ: Lawrence Erlbaum Associates.

Jungk, R., & Mullert, N. (1987). *Future workshops: How to create desirable futures.* London: Institute for Social Inventions.

Kyng, M. (1989). Designing for a dollar a day. *Office, Technology, and People, 4*(2), 157–170.

Nygaard, K., & Bergo, O. T. (1975). Trade unions: New users of research. *Personnel Review 4*(2).

Sjögren, D., & Ehn, P. (1991). From system description to scripts for action. In J. Greenbaum & M. Kyng (Eds.), *Design at work: Cooperative design of computer systems* (pp. 269–280). Hillsdale, NJ: Lawrence Erlbaum Associates.

9 Contextual Inquiry: A Participatory Technique for System Design

Karen Holtzblatt
InContext Enterprises, Inc. Sudbury, MA

Sandra Jones
Digital Equipment Corporation, Nashua, NH

Our goal in system design is to support, extend, and positively transform the work of individuals, teams and businesses through computer systems. We use the term *system* to refer to a complete system that includes hardware, software, services, and support. Contextual inquiry was developed to help achieve this goal and to meet three challenges.

1. We needed to identify a process for designing systems that support people who engage in similar work in many different business contexts and cultures. As Grudin (1990) pointed out, large computing systems are in the business of delivering general purpose computing systems. At Digital Equipment Corporation, engineering does not produce systems for a specific user with whom we can work continuously and to whom we are delivering a system tailored to a specific work process and culture.

2. We needed to identify an expedient process for gathering user information in the time available. Time and resource constraints within engineering and on the part of our customers limit the amount of time we can spend working with users.

3. We needed to identify a means of gaining appropriate and helpful information about users' work. Traditional information collection practices do not seem to provide a context in which users can articulate their requirements in a form that helps engineering design a system that meets our goal.

Contextual inquiry grew out of these organizational challenges. It supports the production of general purpose systems and provides us with a way to work for short periods of time with users at multiple customer sites that are geographically dispersed. It helps people crystallize and articulate their work experience. It also fosters team work and the development of a shared, consistent system vision among team members. (Team members include hardware and software engineers, people and product managers, writers, marketing personnel, and customer support personnel.)

Contextual inquiry is a technique that fosters participatory design. It provides a way for users to participate in the design of general purpose systems. It is a technique for working with users to help them articulate their current work practices, system practices, and associated experiences. The technique contributes to initial design concepts by providing an understanding of the nature of user's work through inquiry with users. We represent this understanding through models of current work practice and descriptions of fundamental work concepts that were developed through interviews with users. This understanding is then used to design a system model that supports user's work.

Throughout the design process, we use contextual inquiry to challenge our current understanding and system design with users. Users participate in iterative codesign through the use of design artifacts, such as work models, paper prototypes,[1] and software prototypes.

Contextual inquiry is an adaptation of field research techniques taken from psychology, anthropology, sociology, and interpretative hermeneutics (Darroch & Silvers, 1982; Glaser & Strauss, 1967; Packer, 1985). Winograd and Flores (1986) also emphasize hermeneutic principles for computer design. The principles and practice presented in this chapter were developed at Digital in 1986. They enjoy increasing use and visibility within Digital and the Human-Computer Interaction community (Bennett, Holtzblatt, Jones, & Wixon, 1990; Good, 1989). Over the past 2 years, we have trained over 400 members of engineering teams in the use of contextual inquiry. This training and subsequent coaching is a key component in our vision to instantiate participatory design processes in Digital.

Contextual inquiry has been combined with quality techniques and decision-making processes to form the backbone of our development of a Contextual Design Software Methodology (Cohen, 1988; Whiteside, Bennett, & Holtzblatt, 1988; Wixon, Holtzblatt, & Knox, 1990). We have also integrated other participatory design techniques, such as mock-ups, scenario building, Metaphor Workshops, and Future Workshops (Blomberg & Henderson, 1990; Ehn, 1988; Ehn &Kyng, 1991; Kensing & Madsen, 1991; Kyng, 1988; Nygaard, 1990) with contextual inquiry. We see contextual inquiry as a set of principles and practices that can be combined with other participatory design techniques to codesign system and system work models with users.

We use contextual inquiry to develop an understanding of current work prac-

---

[1]The work on mock-ups at Aarhaus University (Kyng, 1988) inspired our implementation of paper prototyping.

tice. This understanding becomes the basis for developing a system model that will support user's work. We have combined contextual inquiry with paper prototyping to represent the system model to users in order to involve users in iterative design. When we use a prototype, we inquire into how the future system will support user's work and where the system model disrupts their work. We have also used Metaphor and Future Workshops to help users articulate key concepts in user's work and generate system concepts that will support their work. The principles of contextual inquiry guide our interaction with users regardless of the specific technique or meeting structure that we employ.

A PERSPECTIVE ON USABILITY

We recognize that in order to design a usable system, we must involve users in design. Usability includes the accessibility of the system through interaction with the system interface and the features and structure of the system that supports users' work flow and work activities. As Gentner and Grudin (1990) emphasize, a usable system supports users in achieving their task by hiding the underlying system mechanisms used to support the task. In this section we emphasize the importance of supporting fundamental work process as a key element of usability.

Usability as a Match Between Work Practices and Technology

Fundamentally, usability derives from the optimum match between users' work intentions, concepts, and work flow, and the work expectations that designers build into the system (Jones, 1989). Implicit in every computer system is a model of user work that consists of assumptions about how users work and what they need to support their work. We claim that system models must build on users' existing work intentions, concepts, and work flow—not violate them. The system model must be informed by an understanding of current work practice.

Too often industry develops new technology and then seeks a market for it. To be successful, industry must ensure that technology is supporting a target market. We suggest that an understanding of user's work should be the starting point for defining enabling technology and directing future technology.

We are not suggesting that products replicate work practice without adding new concepts and automating the steps in a process. Rather, we claim that a usable system is one that allows users to adapt their work practice easily and expand their work concepts to take advantage of empowering technology (Ehn, 1988).

The system is perceived as usable insofar as it supports work flow and usage patterns. We might like to think that we can measure usability as one measures the height, length, and width of an object; that by looking at a product we can determine a priori whether it is usable. However, this presumes that usability is an attribute of the system. We claim that it is not. Rather, usability is an attribute of the user's interaction with the system.

Users bring work intentions, work patterns, and behavioral expectations to their interaction with the system. The system provides capabilities that are structured and accessed according to assumptions about user's work structure, work concepts, and patterns of usage. The match between what the system provides and how users want to do their work determines whether the system is experienced as disruptive or usable.

Because much of the quality of this match has to do with how users experience their work and system interaction, we need to involve users in the design of an appropriate system work model and user interface.

Design Distinction: Work-of-the-Work and Work-of-the-Tool

To let an understanding of individual and organizational work drive system definition and design, we must partner with users to get design information and to understand how they work and how they use systems.

The computer system must support different kinds of work, such as system management, software development, or document production. Within and across particular work domains, people have intentions and goals in regard to what they are trying to produce. They decompose and arrange their activities in a particular way, both at a project level and at the level of individual activities.

We have found that people have preexisting work concepts, work language, and work patterns that define their movement through this work. Detailed knowledge about the structure and flow of work domains and tasks allows us to structure the system to support and extend this work. A usable system is one that supports the way people think, talk about, and structure their work (Gould, Boies, & Lewis, 1991).

To do their work, people use tools. If we ask people writing a letter what they are doing, they do not say: "using a pen." The work is letter writing. The tool is the pen. If the pen leaks, it enters awareness as a problem. If the pen is too thick it enters consciousness as a discomfort. If the pen breaks, it enters consciousness because work must now stop. If the pen writes smoothly with flowing ink, it does not usually enter consciousness. The pen becomes transparent; the tool works well (Heidegger, 1962; Holtzblatt, Jones, & Good, 1988).

We make a distinction between the *work-of-the-work* and the *work-of-the-tool*. The work-of-the-work refers to the way people think, talk about, and structure their work. The work-of-the-tool refers to the way people interact with the computer system. When either the work-of-the-work or the work-of-the-tool is violated or poorly supported, the system fails to be usable. Therefore, the kind of information we need for design is information about the work-of-the-work and the work-of-the-tool. Design information is detailed information about what supports, what hinders and what could positively transform the work-of-the-work and the work-of-the-tool.

Traditional usability work has concentrated on improving the work-of-the-tool. People approach a system with preexisting tool patterns that can be built on or violated. The ease, comfort, and fit of the tool is the result of how well the designers[2] understand what supports and what hinders tool use. A concentration on tool use tends to place usability work at the end of the development cycle. A dual concentration on tool use and work process moves usability work forward in the development cycle.

The contextual inquiry process brings engineers, designers, managers, and marketing personnel to users to acquire design information. We claim that when designers are called on to make decisions in the course of product development, they do so from an understanding of users' work. Through contextual inquiry with users, designers can design from an understanding of users' work.

PRINCIPLES OF THE CONTEXTUAL INQUIRY PROCESS

Contextual inquiry does not provide a set of steps to follow for collecting and interpreting user information. Rather, it provides a set of concepts that guide the design and implementation of information collection and analysis sessions. Three principles guide the contextual inquiry process: context, partnership, and focus.

Context

We have claimed that the definition of a system that supports and positively transforms users' work is derived from an understanding of users' work. Understanding users' work is, therefore, critical to system design.

The best way to understand the work is to talk to people in their actual work environment. Design information is present in its richest form when we speak with people during ongoing work or using work artifacts. Work artifacts are the results of work. For instance, the artifacts that an application developer generates usually include functional specifications and code modules.

---

[2]We use *designer* to refer to the person who is talking with users about their work while thinking about the system design. This person can be anyone on a cross-functional team, such as an engineer, a product manager or a human factors person. This is not meant to exclude the role of co-designer that the user takes on during the participatory design process.

People Speak About Their Work in Abstractions.   Typical data collection techniques, like surveys, structured interviews, or focus groups, collect information about what people do or what they want in a new system.

The notion of collecting system requirements suggests that people have the information that we need to design a system. If we ask people what they *want* they will be able to tell us. And, if we collect all this information we can then design a system to meet users' wants. This assumes that people are aware of what they are doing, and know what they need to improve the system.

We claim that people do not usually think about their work process and the impact of computer tools on their work as they work. People are engaged in doing work; they are not simultaneously reflecting upon their experience of doing work. Our focus of attention throughout the day depends on our particular concerns. For example, people who are building aircraft do not simultaneously think about the design of the computer tools they are using. The perspective from which computer designers observe users' work differs from the users' perspective.

We have found that when we talk to people about their work in an informal way or in traditional interviews, they tend to speak in abstractions and summaries about what they do and how they work. People have typical tasks that they can talk about. If we ask them to describe these tasks, they speak in abstractions about typical meetings, what usually happens, a philosophy of work, and other musings about their roles. This alone is not information that we can use to design a system.

On a visit to one of Digital's customers, we asked a user to describe the process for application development. The user started to describe the roles of various people and then showed us volumes of documentation that describe their processes. He said, "This is the way we work." We then asked about his current project. We asked whether he used the processes described in the volumes for his current project. "Well, no. There were extenuating circumstances and resource issues. Parallel development was needed." When we asked about his previous project, we got the same answer.

We find that process descriptions are idealized models of what should be or could be a way of working, but they rarely describe the actual work flow of people trying to accomplish a task. Similarly, descriptions of personal work processes are abstractions of general and common features of work derived from many instances of work, each with its own particular character.

The farther away we are from actual work, the more abstract the descriptions of work become. If we design a system based on these abstractions, we will likely fail because we do not know the actual ways that people get things done.

Through contextual inquiry, which focuses people on articulating their work experience as they work, we can help people to be concrete. Through dialogue with users, we develop an understanding of their everyday work flow by focusing them on particular instances of work. We can then create a system work model that supports users' work.

The Difference Between Summary Information and Ongoing Experience.   When asked about a computer system, people provide a summary statement, or their opinion of the system. This summary statement is an abstraction derived from everyday experience. For example, when a writer was asked what she liked about the text editor she used, she said:

> I like it because it was really easy to learn. It uses this keypad with the labeled keys so I don't have to remember what key to press. I also like the multiple buffers.

She liked the multiple buffers feature in the editor. Frequently, people are not aware of what they like about a system, because it works well and does not call attention to itself. Or, they might tell us what features they like, but not sufficient detail about the feature and how it supports their work flow.

If we are present when the person is actually using the system, we can capture the detail of ongoing experience. For instance, during the contextual interview when the writer was editing a document, she began using multiple buffers, saying:

> This is a really usable feature. I position myself in one buffer, get the buffer I want to select from, return to the first and just pop it in—the cursor holds my place. I love it! It's a great productivity gain.

The ability to cut and paste between buffers easily was central to the usability of multiple buffers. Editing between buffers is usable because the editor maintains the last cursor location in each buffer. Multiple buffers without cursor placeholding would not be as usable. Similarly, when asked what she disliked about the editor, she said:

> Well I can't think of anything. There are little things that get in the way on a daily basis, but right now I don't know what they are.

Such summary information does not provide us with insights about what is "in the way." However, during the contextual interview she pointed out a problem with the way the editor failed to handle text formatting. The editor did not automatically wrap text as text was being edited. This prevented her from reading all of the sentence as she edited. She spoke of the disruption this caused:

> This is a biggy. I wish they would fix this.

Outside of the context of actually doing work with the editor, she was unable to recall this problem. It was the experience of encountering the problem that brought it to memory. In fact, she mentioned the problem before it actually occurred on the screen. As an experienced user of the editor, she had developed a workaround to the problem, a way of avoiding the situation. As she was about to use the workaround, she remembered the problem and discussed it. Even though the problem did not occur, the disruption to her work did occur.

Only by being present when work is happening can we see and discuss the details of what supports or what hinders work. When people are not in the context of actual work, they are not aware of the details. If something is very disruptive and causes work problems, it stands out in their memory. Little things build up over time. People might not know what the problems are; they only know that they do not like the system. On rating scales they rate the system low, on open-ended questions they list major problem areas to improve, and at interviews they discuss *bug* lists and *wish* lists (their design solutions) saying: "Support me; fix the system."

If we ask people to tell us what they do, what works for them, what problems they have with a system and how they work with others, we get summary experience. It is likely that we will, at best, collect attitudes about the system, abstract descriptions of work and lists of very disruptive problems. None of this *information* tells us what to design to support the work. Moreover, people do not naturally conceptualize their work, the work process, or the elements of design that support or disrupt that work. These are not users' major concerns; their major concern is doing the work. However, these are the computer designer's major concerns, which can be discovered by talking with users about their work while they work.

Accessing Ongoing Experience.   Being present in the work context leads to more information than other kinds of information gathering. The workplace is filled with *touchstones* that help people recall their work experience: books and documentation that are open or unwrapped, crib sheets taped to the walls or computer monitors, documents used in meetings, files stored in a structure that supports the work, the coworker down the hall who is the source of problem solving, office mates, the broken door on the workstation from frustrated rebooting, the calendar with appointments, and the artifacts that people produce.

Walking with people through their work place and focusing them on their concrete experiences produces concrete information. People see something that reminds them of what worked and what did not work, and of stories about how things happened on the last project. The things in the environment provide the designer with opportunities to ask how objects are used now or how artifacts were produced. The people in the environment become the source of conversation about teamwork and organizational culture. The set-up of the environment, office layout, dress and degrees of neatness, lead to conversations about structure in the organization, interpersonal work, and expectations.

Contextual inquiry focuses people on their work and tool experience in the context of actual, ongoing work. Being present while the person works with a tool or with others allows the designer to witness the person's work and system experience as it occurs. Through the use of paper prototyping in actual work contexts users can imagine the effect of a potential system design on their work. Through dialogue with users in their work context, we can design a usable system work model and user interface.

Partnership

To design effective systems, we need to understand users' *experience* of work and systems. This information is invisible; we cannot access it by standing on the outside of a process, watching people's behavior and writing down what happens. We need to talk with users to understand their experience. To have an effective dialogue, we form partnerships with our users.

The principle of partnership recognizes that only through dialogue can designers become aware of users' experience of work and tool use. Together designers and users create a shared understanding of work practice that reveals technological opportunities and problems that occur in work processes and in system use. From this shared understanding designers create usable systems.

The User is the Expert.   We recognize that in the area of personal experience and work, users are the experts. As such, users must act as informants in their relationship with designers. Designers do not know the users' experience; users must speak their experiences.

Acknowledging users as the experts and the source of understanding has two important results:

First, when we say that users know and the designer does not know, it becomes acceptable for the designer to ask questions. The designer's role is not to solve problems or give answers. Rather, the designer's role is to understand users' work and experience in order to imagine technological solutions. Since users know, and we do not, we can free ourselves from the expectation that we should already know and let ourselves inquire about what is going on.

Second, acknowledging that users know, protects us from misinterpretation. Interpreting another's behavior into our own experience is a normal occurrence. We do not look at a behavior like a smile and say only that Joe is smiling. We usually assume something about Joe's inner experience. We say that Joe is happy. We also look at the context and assume something about the cause. Joe is happy because Mary smiled at him. We only witnessed behaviors. Yet, we derived meaning from these observations. However, our interpretation may be incorrect.

We interpret users' behavior based on our experience and previous knowledge of the work context and tools. When we observe behavior, we impose our interpretation on the person's intention and experience. When we make assumptions about "what is going on," we can be wrong. Through dialogue we let users shape our assumptions and the meaning that we derive from a situation.

For example, during an interview we thought it annoyed the person to have to press several keys to correct a typographical error. When we asked the person about our interpretation of her behavior, she said that correcting the error was "like one keystroke." The layout of the function keys allowed her to correct common errors quickly with little disruption to her work. Acknowledging that users are the experts frees us to ask questions, to let go of the need to provide answers to technical problems, and ensures that we shape our interpretations by the users' experience.

Sharing Control During the Inquiry.   In a traditional interview or survey, the "asker" generally has the power. The asker arrives with a list of questions. In this setting, users understand that their role is to answer the questions. The asker controls the pace and content of the conversation. Generally, one of the unspoken rules is that users do not question the questions. Users answer the questions.

But we have said that users have all the information, including what is important for us to know about. We could form a list of questions that comprise our best guess about what is important, but we might be wrong. We are not users, and we do not understand their work context. The purpose of going to the user is to find out what is important. If we come in with a question asking format, it implies that we have already decided what is important even if we later ask: "What have I left out?" Because we are the designers, users might conclude that we already know what we need to find out.

To create equal dialogue in which both people create and direct the conversation, the designer must deliberately let go of control and create a situation that allows users to share control. Partnership means that we empower users to direct the conversation along with the designer.

We give up control by giving up the role of expert, by using open-ended questions that invite users to talk, and by letting users lead the conversation. Rather than entering with a list of questions, we enter with a few areas of concern (the focus) which we share with users. This communicates the scope of the conversation. Alternatively, the conversation can begin with a request to: "Tell me about your work and your current project." The dialogue that takes place while users are working and involves probing what is going on. "What are you doing now?" "Is that what you expected to happen?" When we are talking about artifacts, we ask how, why, and when the artifact was created and used. In most cases, the central question brought to the inquiry is "What is going on here?"

The flow and pace of the conversation tell us whether we have formed a partnership of shared control. If the conversation is jerky (stop and go), if the person waits for us to ask a question, we have probably generated an "I ask; you answer" situation. When we are in dialogue together, the inquiry flows, ideas arise, and the interaction is conversational.

Often we enter the interview with a concern about covering all the issues with every person. Since each person's experience is unique, each dialogue is unique. Users cover the issues that are important to their work. We need not cover every issue with every person. There are additional users from whom we can get information. During the interview, the inquiry evolves. No two people are ever treated alike because we intend to articulate their unique experience.

Creating Shared Meaning.   It is often the case that we do not actively listen and build on what others say. Consider a typical meeting: Person A makes a point. Person B thinks of a response. Person C gets his point in first while Person B develops his idea. Person D makes his point. Person B looks for an opening. Finding a pause in the conversation, Person B makes his point. However, the point does not follow from what C or D said, and B barely knows what they said because he was busy preparing his comment. The result is not shared meaning, but individuals getting in their points.

In this case, person B was self-listening—paying attention to his own thoughts rather than listening to others. When we self-listen, we develop our own interpretations rather than a shared interpretation with another.

Tape recorders are useful for capturing words, but they do not necessarily capture shared meaning. What will these words mean later on? If we have not actively created meaning during the interview, we have to impose an interpretation on the words after the interview. Each time we stop listening to users we fall out of partnership and fail to create meaning together.

To maintain the partnership, the designer must stop or change the conversation when either person seems unengaged. Self-listening naturally happens. We get design ideas. When we design in our head while users are talking, we are no longer letting the user shape our understanding. But if we simply share our design ideas as they occur, we both remain in partnership and we can codesign with users. In this way, users' experience of work can be directly translated into system possibilities. By sharing design ideas and interpretations as they occur, ideas get quickly shaped and validated, and the user partnership is maintained.

Reflection and Engagement.   The key to partnership is to maintain a conversation that allows users and the designer to reflect on, question, and create shared meaning about the work experience. This process is one of oscillation between engagement and reflection. Engagement occurs when we actively listen and speak with users about their ongoing experience of work. Active engagement has the sense of a stream of consciousness discussion. Reflection occurs when we stop to consider and integrate information into our evolving understanding. Together we step back to make sense of the ongoing experience. Then, we return to the stream of conversation while users continue working.

We maintain an equal partnership by allowing both people to steer the conversation. Together we create an understanding of work and design ideas that can support the work. Maintaining our relationship to users and balanced control is key to the principle of partnership. Partnership creates the opening for participatory design.

Focus

Focus is the way we manage the conversation with the user. Focus creates and directs our developing understanding. A focus is the filter through which we interpret the world. We perceive the world through a filter comprised of our assumptions, beliefs, and concerns. As a result, we tend to see those things that we care about and ignore those things that we do not care about.

We can learn about focus by considering what happens when we are about to meet a new person. For example, if a friend tells you that Joe is arrogant, you now have an arrogant filter through which you see Joe. Anything that Joe does that could be seen as arrogant is taken as evidence that Joe is arrogant. Anything that disconfirms arrogance is often ignored and attributed to the situation. Even if we resist seeing Joe as arrogant, we are likely to see him through a "not-arrogant" filter or lens. Nonarrogant behavior is magnified and arrogant behavior is minimized. The lens that we apply when looking at Joe's behavior fundamentally influences our interpretation of Joe.

Similarly, we interpret what we see as we work with users through a filter. Depending on the filter we look through, the world looks different. Contextual inquiry acknowledges the role of focus and provides a framework for acting responsibly with our focus.

Fixing the Focus

The designer begins any inquiry or system design with a set of assumptions. These assumptions concern the nature of work, opinions about design ideas, beliefs about usability, and goals for the inquiry itself. The designer begins with an existing understanding of users' work and system use. This is the designer's entering focus.

Once, we showed a videotape of a contextual interview to a group of industrial designers. At one point, a viewer exclaimed: "Look at that. That is terrible! That workstation background is black and the contrast is horrible." This person's knowledge and expertise provided a filter through which she saw the tape. Her focus included attending to the color and contrast on the screen. Numerous other people had seen this tape and no one ever commented on the screen characteristics.

When people with different backgrounds see the same situation, different aspects of that situation are revealed. Similarly when we fix, shift, and expand our focus we can systematically alter what we attend to in an inquiry.

Having a particular entering focus directs information collection and analysis. This focus determines the goals of a study, what is attended to or ignored, what questions to ask, and what to probe further. When we enter a situation without a clear focus, our attention is random; we are uncertain of our purpose in the conversation and we do not know what to follow up. This uncertainty causes some discomfort with open-ended interviews that give users control of the content of the conversation. We need to prepare or articulate our focus before the interview by clarifying the domain of work that we want to explore and by defining our design problem.

When we know our entering focus, we attend to and probe relevant aspects of the conversation. If we lose track of our focus, we can easily recapture it by looking at our notes defining the focus. Rather than trying to eliminate focus, we use it to give us richer information. Sharing our focus with the user provides a way for the user to codirect the conversation.

Expanding the Focus.   We have said that the purpose of the inquiry is to uncover users' experience of work and systems. We usually enter the inquiry with a set of assumptions about work and system use that may reflect a small slice of the users' experience, and a set of assumptions that are not appropriate. To bring our understanding in line with the users' experience we must expand and shift our focus as well as our understanding of users' work and system use. Focus is dynamic. To create a shared understanding of users' work and system experiences beyond our entering focus, we must probe their words, question their actions, and share our ongoing interpretations. Figure 9.1 depicts the expanding focus.

During each interview and as we go from user to user, we build an understanding of users' work and system usage. As with any understanding, we begin with a structure into which we fit new information. This understanding develops and changes as we incorporate the new information. Only when the evolving understanding can no longer incorporate the new information, no matter how we twist it, do we shift our paradigm (Kuhn, 1962). In contextual inquiry, we expand the focus deliberately to speed the process of changing our understanding in response to new information.

To expand our focus we must deliberately attend to that which we would normally ignore. The typical human response (consciously or unconsciously) is to ignore information that does not match our expectations. We can expand our focus by attending to surprises or contradictions. Any time we are aware of a contradiction between what we are seeing and what we believe, we deliberately

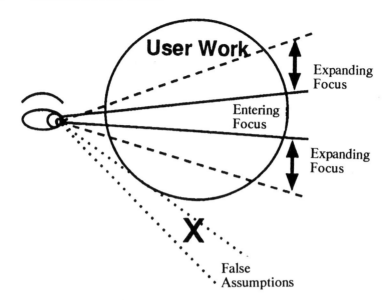

FIG. 9.1.   Expanding the focus.

explore the situation with users. Instead of dismissing information as pertinent to only one user, we probe into the contradiction by asking the person to explain the situation that surprises us. We expand our focus by continuously challenging it with new information and asking what better construction can represent the experience. This is a fundamental commitment both in the inquiry itself and in the overall design process.

To expand our focus we ask about what we do not know. If we are confused about the technology or overwhelmed with the information, we might think that it is best to let the tape recorder pick up the information and ask someone knowledgeable when we get home. When we do this, we lose our partnership with users and we do not understand what is going on any longer. Our understanding can only expand when we continuously integrate new information. So to expand the focus we simply ask people to explain. As they explain, they will elaborate and provide more information. We remember that they are the experts; we are not expected to know.

Checking with the person about what we think we understand is another way to expand the focus. We call this probing the "nods." When we have had similar experience, we hear what users say as "known and understood" because it is "like my experience." We need to be careful to avoid assimilating users' experience into ours; we could be wrong. Because we often nod our head when we think we know something, we use this cue to remind us to probe. We tell the person what we are thinking so that we can validate our interpretation as correct. This also provides an opening for users to provide additional information.

We expand our focus by probing behind solutions that users offer us. When working with existing systems, users provide us with their design solutions to their problems. These wish lists hide the work or system problem for which the wish is a solution. Only by understanding the issues behind the solution can we collect knowledge of the impact of the system on users' work. When we see the whole work pattern, we can redesign a system to support the work. To expand our focus, we probe behind the solutions offered to uncover the origin of the problem.

Finally, we expand our focus by sharing our interpretations and design ideas. By sharing our interpretations and ideas as they unfold we can get immediate feedback and elaboration from users. Users might say: "Yes, that is just what I meant" and go on to elaborate in greater detail. Or they may look strangely at us and say: "No, I didn't mean that at all." In the latter case, we can go on to explore further what they did mean. In this way users shape our understanding to ensure that it conforms to their experience.

Interrupting users in the moment allows them to stop and reflect on their experience at the time of their experience. We recognize that we cannot get information if we do not interrupt. Yet we need to interrupt carefully, waiting for users to complete an action and paying attention to the frequency of interruptions. As the process of reflection and engagement continues, the inquiry feels like a conversation, not a series of interruptions.

If during the interview, an assumption that we have does not arise, we can discuss it at the end of the interview during the wrap-up. During the wrap-up, we pose an assumption to users and listen to their response. People may want to agree with us just to be nice. Moderate statements of confirmation cannot be considered a confirmation. Gauge the intensity of the response. If users say, "Oh, how could I have forgotten to tell you about that!" and go on to elaborate, we are receiving confirmation of a work assumption or design idea. Conversely, if users respond with little enthusiasm, with confusion or rejection, the assumption is disconfirmed.

To expand focus and ensure that our understanding of work and system usage represents the users' experience, we probe the users' words and ask questions about their actions. We probe to obtain information relevant to our entering focus and we probe to expand our focus. We share our interpretations and design ideas. Checking our interpretations both during the inquiry and at the end of the session helps to expand our focus and eliminate unsupported beliefs during the interview. Through our changing focus, previous designs are discarded and new designs emerge. As we go from user to user and from business to business our understanding of the work domain broadens, a system work model is developed, and a user interface design is defined. In this way, multiple users codesign a generic system.

The Technology Focus.   We have said that a new system vision comes out of an understanding of users' work. But how does this happen? Design is always a matter of seeing the possibilities within constraints. Just as an interior designer's challenge is to create the desired look using one's old furniture within the structure of the house and for the budget available, a system designer is constrained by the work of users, the technical possibilities, time, budget, and resources. Our challenge is to build a system that supports users' work given the available technology and other constraints.

Like any focus, the technology focus reveals and conceals. It shows the designer what is possible. However, it conceals what is outside of it and is limited by the designer's current understanding of the technology. Technologists can dismiss ideas as "not possible" when implementation is not obvious. Unfortunately, if we think only about what can be implemented today, we miss the possibilities for a creative system and for technology development. Most importantly we can miss new ways to combine existing technology.

Behind "can't do" thinking often lurks a fear that if something is discussed, thought about, and written down there is an implied commitment to build. For this reason all of our processes deliberately separate possibilities thinking from decision making. We try to emphasize to designers and engineers: "Just because you think something, you don't have to build it." Once we look at the possibilities, we can look again with a critical eye toward feasibility within a given time frame, resource allocation and market window. If we shut down creative possibilities before they are developed, we might eliminate a workable solution prematurely.

By sharing our technology focus with users, users participate in envisioning a system that can transform their work. Users and designers represent the system vision by creating paper prototypes.

CONDUCTING A CONTEXTUAL INTERVIEW

A contextual interview is a manifestation of the principles of contextual inquiry. In this section, we describe the setup that is required and the structure of an interview. We include practical hints based on our experience conducting interviews.

Identifying Customers

We begin the interview process by asking: Who is the customer? We think of the customer in two ways: as the business or industry that we expect to buy our product and as the individuals interacting with or getting value from the system. We work with marketing and sales representatives to determine the market for the product. From this discussion, we generate a list of key industries. Because we would like to return to users throughout the development cycle, we try to establish customer partnerships with key customers. Key customers include industries that are critical to our business success that engage in the kind of work the system will support. These may not be the only customers we work with, but they are a core group that represents a good cross section of the marketplace. Once we determine a set of potential customer partners, we work through business channels to contact user representatives.

Arranging the Visit

Before contacting the customer, we develop a framework for our visit that includes our purpose, a structure for the day of the visit, specific requests regarding the roles of users we want to interview and the amount of time the interview will require. The framework helps us to talk with user representatives about what we would like to do. We vary it according to specific situations. We also discuss our intended use of audio or video recording and send a copy of the form that we use to procure written permission before the visit.

A typical day at a customer site might look like the following:

- 9:00–9:30
  Interviewers meet group of customers to describe the product area and the process and to pair off. This group may include management and other contact people.
- 9:30–12:00
  Contextual interviews in offices or work place.
- 12:00–1:00 Lunch as a group.
- 1:00–3:30
  Second round of contextual interviews with a second set of users.
- 3:30–4:00
  Team consolidation and quick review of interviews if possible.
- 4:00–4:30
  Discussion with hosts about the results including a summary of major understandings.

We send thank-you notes after the visit.

Selecting Initial Users

Once we have determined what users we want to work with, our focus determines the particular individuals we want to interview. In determining who to interview, we consider the following roles:

- People who use the system directly (users),
- People who manage the users,
- People who receive products from the system (recipients),
- People who test the system and provide purchasing recommendations to management,
- People who make the purchasing decision, and
- People who use competitive products.

In a large work domain, our users have many different kinds of work. For example, the work domain of system management leads us to talk with operators, system managers, managers of the group, network managers, PC network managers, hotline operators, and the decision-makers. The team needs to identify the kinds of work that will be done with the planned system and who will be using the system. They can then identify people in key roles to be interviewed at a single site. When each interviewer works with a person doing a different type of work, the team will gather a sense of the business, culture, team interaction, and particular work when the information is combined.

We have found that it is best to schedule one visit at a time. This allows time between visits to analyze the information and refocus for the next set of interviews. We want to be in a position where we can make decisions about who to interview next based on our evolving focus rather than on an a priori decision.

Selecting Subsequent Users.   Focus directs the interview; focus also directs the decision of who to talk with next. We use the time between visits to synthesize the information and frame our unanswered questions and concerns. This understanding guides us in selecting users for the next interviews.

Through the interview process, we are seeking to expand our focus and make it consistent with the users' experience. As such, we look for mismatches with what we already know; we are not trying to replicate experience. We select users who we think will challenge and develop our understanding. We are less concerned about replication than about whether our current understanding remains valid as we move from person to person, industry to industry, team to team, and across different kinds of work. To create a generic system we are looking for relationships between fundamental elements of the work domain. In addition, we attend to site-specific information that will tell us how to design customization and preference options in the system.

We are often asked how many people we need to talk with before beginning design. We accept that within engineering we are always designing from partial understanding. When we have no user design information, we are designing from an n of 1, where 1 is the designer. We recognize that we can never talk to *enough* people to ensure completeness for a generic population. As we continuously expand the base of users we work with, our understanding expands to respond to a larger segment of the targeted market.

Using Multiple Interviewers

Multiple interviewers contribute to expanding the team's focus and the understanding of users' work issues. There are several reasons for using multiple interviewers:

1. If a team collects information from several users within a project at a site, the information collected will reveal how people work together and the flow of work in the organization. This is a way of seeing a cross-section of a customer's business.

2. Interviewers with different functions, such as documentation, engineering, product management, training, and support, have different perspectives. Although the interviewers are working from the same focus, each person sees things from the perspective of their expertise. This difference is a strength. Through team analysis, the different perspectives are incorporated into a system vision. Interviewers with different functions can also incorporate each other's perspective in future interviews.

3. Interviewers with the same function, such as engineers, also have different perspectives. Each person sees things differently. Through team analysis, each member comes to understand and incorporate other perspectives into the system vision.

We deliberately develop cross-functional teams to conduct interviews. We bring small teams of three to five people to a customer site. Usually, each person conducts two interviews with different users within one day.

We have also conducted interviews in pairs or small teams. The advantage of this approach is that the team develops a shared understanding during the interview. The disadvantage is that it reduces the number of people we can interview during a visit.

Setting the Focus Before the Interviews

Before going to the interview site, the team of interviewers meets to articulate their focus for this visit. If the team will be talking to people who do different types of work, the group discusses what aspects of that work will be probed. The team deliberately articulates its entering assumptions and focus, making notes on key areas identified from previous information, and clarifying its purpose. The interviewer discusses these notes with the user to focus the conversation. These notes can be referred to during the interview to regain focus if the conversation becomes undirected.

Structure of a Contextual Interview

A typical contextual interview has three parts described in this section:

- Introduction: Establishing a relationship
- Ongoing work inquiry
- The wrap-up

During the interview we take notes. In our notes, we write down:

- What users do
- What users say
- Our interpretations
- Disruptions that occur to the users' work including workarounds
- Aspects of tool use that support the users' work

Our focus directs what we discuss. During the interview, we integrate information continuously into an understanding of users' work, and share the understanding with users. We do not try to structure or categorize the information during the interview.

It is more important to maintain the inquiry with the user than to take notes. However, when we are in the middle of capturing a thought and users move to the next point, we ask them to wait until we are finished recording. If we are unable to use recording equipment, we might work in pairs and have a colleague take notes to supplement our notes.

Introduction: Establishing a Relationship.   A contextual interview begins by establishing a relationship. We introduce ourselves to users, tell them why we are there, assure them that all they say is confidential, and ask for their permission to record the interview. (We might use written consent forms when using recording equipment.) We tell them that we are interested in learning about their work process and how they experience their work and the tools they use. We enlist them as partners in the inquiry by telling them that we are there to learn about their work and use of computer tools.

We tell people how long we intend to be there, and that they can take a break or end the interview at any time. We ask them to give us an overview of their work and to discuss what specific work they will be doing during our visit. We ask for their opinions of the tools they use.

People always have things they want to tell us, such as software *bugs* or *wish* list items. Sometimes they want to demonstrate software. We listen to what they say. This conversation gives us hints about what to look for during the ongoing work part of the interview. Until they tell us what they need to tell us, they cannot focus on doing their ongoing work.

Ongoing Work Inquiry.   As we move into the ongoing work part of the interview, we ask questions and share ideas as users work. We tell users that we will be interrupting, and ask them to tell us if we interrupt them at an inconvenient time. When we do interrupt at an inconvenient time, we make a note of our thought or question and ask it later.

As users work, we do not talk all the time; we also watch in silence. In fact, there are often long periods of quiet. The objective is to build a shared understanding of users' work. To do this requires discussion, but we do not sit there generating questions. We are engaged in an inquiry with the user. Our work is to maintain that partnership. We do this by sharing our interpretations and design ideas.

After we have been assimilating information continuously we may reach an overload point and need to go away and put our ideas together. We either suggest a break or that we end the session. (It is often the case that the interviewer tires before the users.) If we are tired, we say so and end the conversation.

We need to ensure that users are equally involved in the partnership. The role expectations often cause users to avoid requesting the session stop. Therefore we attend to any nonverbal and subtle cues (looking at watches, mentioning hunger) that suggest that users need a break or want to end, and suggest this.

Our goal is to articulate a coherent understanding of the user's work process, uncover the needs for the work, and uncover what supports or hinders the work. We let users lead the discussion by keeping the questions open-ended. We need to be careful not to narrow our focus too quickly, and try to keep specific questions until the wrap-up. We probe behind solutions and follow-up comments that contradict our entering assumptions. By so doing, we expand our focus and co-interpret the work with users. If users go off on a track that is not useful, we redirect the conversation.

We stay concrete and keep people concrete by attending to the ongoing work and work artifacts. For example, if they say, "People here do this," we ask: "What do you do?" If they refer to a "DFD," we ask them to explain the acronym and to show us one. We speak from concrete artifacts, ongoing work, or recently finished projects. When they refer to something, we ask to see examples.

When people are using computer tools for their work, we ask to see them use these tools. Some questions that we ask to keep the conversation going are:

What are you doing?
Why are you doing that?
Is that what you expected?

We ask them to tell us about workarounds that they have developed to avoid problems. Probing workarounds reveals how people think about and structure their work.

We bring out assumptions that we have (our focus) when people refer to that area to see if what we thought matches what they think. This is an opportunity to validate or invalidate our assumptions. Similarly, we share our interpretations and design ideas for validation:

> I'm hearing you say such and such. Am I on track?
> What you're saying leads me to conclude . . .
> What if the application did . . . ?

The Wrap-Up.   To conclude the interview, we summarize what we learned by referring to our notes. This is a good time to clarify any uncertainties. If particular questions or concerns were not raised during the interview, we discuss them at this time. We need to be careful about interpreting users' response; just to be polite they might say that they like an idea. This is an important time to validate or invalidate our assumptions. Finally, if users are using a product with which we have experience, we give them tips that help them be more efficient.

We ask if we can call later if we have additional questions, and invite them to contact us if they think of something after we leave. We close by thanking them for their time.

Variations in the Use of Contextual Inquiry

The principles of contextual inquiry direct us to collect information about work and system use from people as they work. Within Digital, it is not practical to spend days, weeks, or months with users. Therefore, we seek to maximize our information by spending several hours watching and talking with people as they work. Typically we spend two to three hours and we may make return visits. However, not all work can be seen in a day or less, and not all work can be interrupted. We use the principles of context (closeness in time, concrete instances, and accessing ongoing experience) to design alternative information collection schemes.

For example, application software development takes place over months and years. To get the flavor of this work within 2 hours, we take a team of three to five people to the customer site to interview different people performing different kinds of work on the same project. This gives us a sense of the way work overlaps and is coordinated. We also have people speak from their current project in the context of the artifacts that they produce and tools that they use. This allows us to talk with users about work process in a concrete way within a limited period of time.

For work that is difficult to interrupt, we might videotape a work session and immediately review the tape with the person in order to interpret what happened. The videotape is used to evoke the experience that the person had in the moment. In this and other ways, the principles of contextual inquiry can guide the design of information collection sessions.

ANALYZING CONTEXTUAL INQUIRY INFORMATION

Analyzing contextual inquiry information is interpretive and results in a shared understanding of users' work and system use that informs design. Throughout the process we are building an understanding of user and organizational work practices, a system vision, and specific design ideas. We are also generating questions for technology and future users. Analysis takes place both during and after the interviews.

During the interview, we engage in simultaneous information gathering and analysis. We share design ideas that are pertinent to the work being discussed and we engage users in design conversations. Together with users, we cointerpret their experience of work and usability. At the end of each interview, we have an understanding of their work and system usage.

The process of postinterview analysis is similar to the interview process. However, rather than having a dialogue with the person we dialogue with the text by reading the transcript and asking questions from a particular focus. Our interpretation continues as we analyze the text. To date we have not had users present during postinterview analysis. However, we have generated questions for clarification and followed-up with previous users.

The team performs additional analysis after the interview. We recommend team analysis over individual analysis and will describe a process that uses Affinity diagramming to support team analysis. Affinity diagramming is a technique for building a structure of ideas from a set of unstructured ideas. It is an inductive way to conceptually group information (Cohen, 1988).

Analysis by a team has many benefits. It integrates multiple perspectives and creates a shared vision of the system and a shared focus for subsequent interviews. It builds team spirit and saves time in the long run. We have found that individual analysis used by others results in the individual compiling documents that are neither read nor convincing. The individual ends up explaining conclusions and team time is used to redo what was done alone. Although the logistics of working as a team can be difficult, in the long run working as a team saves time. Team analysis also contributes to group ownership and builds a clear understanding of the whole system vision.

The following description of analysis focuses on a team process. Readers can abstract principles and apply them to individual analysis situations.

Transcribing the Interview

It is important to transcribe notes and tapes while the interview is still fresh in memory. Reviewing the interview helps us refocus for the next set of interviews. As we create the transcript we may include comments, insights, or questions that arise, using notation that differentiates our comments from the interview content.

Transcription can be time consuming (we estimate a 2–1 or 3–1 ratio for audiotape transcription). We have had secretaries and a transcription service transcribe the tapes so as to minimize the time team members spend on this activity. It is much easier for the team to work from a transcript or typed notes than from the original tapes or handwritten notes. At the beginning of the transcript, we include an orientation to the user's work and any other relevant demographic information.

Fixing and Evolving the Focus of Analysis

It is important to clarify our focus before beginning analysis. When we are doing analysis with a team, the team clarifies their focus together. We ask: What is our purpose or design concern? Without a clear focus, we can get lost in the information. Our focus directs what we include and what we exclude.

If we were developing a tool for debugging software programs, for example, our focus could be: "What is the work of debugging?" "How do the tools that users have support or disrupt the work of debugging?" As we read the transcript, we look for responses to these questions. After reading a section of text, we ask: "What is going on here?" "What does this tell us about the work of debugging?" We exclude information that is not about debugging. If we changed our focus, we would include different information.

Interpretation results from this dialogue between our questions and the transcript. We read and reuse the person's language, moving back and forth between specific instances and the entire session. As we interpret the text, questions may arise. In fact, the postinterview analysis may generate an interpretation that needs to be validated either with the person who was interviewed or with a new person. These questions, together with our current understanding of the users' work, form our entering focus for subsequent interviews. Before the next visit the team reviews these questions as part of setting their entering focus.

Analysis of the transcripts has the same effect as analysis during the interview: Our understanding of users' work changes and our focus expands and is grounded, inductively derived from the user information.

Interpreting the Information

The focus that we bring to the information guides our interpretation. As we change our focus, different aspects are revealed in the information. The deliberate shifting of focus provides a rich understanding. As a result, the same information can be analyzed with different foci revealing different things.

The transcript is used as a catalyst to direct team discussions about the user's work and the system. We interpret the text by alternating between reading the text silently and then sharing design ideas, questions, and work observations, and identifying technical issues. If more than one person has conducted interviews, information from one interview reminds us of information from another. This related information is shared to build an understanding of how people work together and the relationship between different types of work.

As we interpret the information, we do not critique ideas. We include ideas that seem technologically impossible or impractical. We keep the same attitude that we brought to the interview; we imagine the impossible. We do not want to constrain our creative ideas. There is no commitment to action at this point; commitment comes later.

We use and reuse the user's language, listening to the user's language carefully. What are users saying? We quote users verbally and in writing to discern the meaning which might have gone unattended during the interview. We consciously resist the temptation to rephrase their words in a way that supports our agenda items. As we move back and forth between the specific instance being examined and the whole session, we develop an integrated understanding of user work. As we understand the work, design ideas naturally emerge. We record design ideas, system work models, and user interface ideas.

The understanding that we develop is specific to the particular context within which we gathered the information. However, the knowledge is also reusable; it is carried across interviews. The knowledge that we develop from analysis becomes our entering focus for subsequent interviews. In addition, some of the knowledge extends beyond the specific context. Knowledge about work structure, work concepts, and usability concepts apply in other contexts. In this way, the understanding that we develop acts as a theory of work and theory of usability that transcends a particular domain. The concept of disruption is an example of a reusable concept that was articulated through analysis of contextual inquiry information. Holtzblatt et al. (1988) found instances of the system disrupting the work, such as when the writer had a problem with text not wrapping in the editor. We now consciously look for and discuss disruption by the system with users in a variety of work domains.

This reusable knowledge provides us with a framework or "hooks" that we use whenever we do interviews. These hooks include attending to:

- Work structure or work flow
- Problems accomplishing the work
- Problems in system use
- Disruptions caused by the system
- Workarounds that are used to avoid disruption from the system

- Transparency of the system
- Aspects of work process and system use that support work

Recording Understandings

As we interpret the text, we record our understandings. This can be done in the transcript by using a coding scheme to distinguish between the text of the interview and the postinterview interpretations. Another way of recording understanding is by using note cards or Post-its.[3] When working alone or in a team we have found Post-its an effective medium as they allow us to group and regroup information easily with the Affinity diagramming process.

When working in a team, everyone has a copy of the interview transcript. The person who conducted the interview begins by giving an overview of the interview and talking about the user's work and background. The team reviews the transcript together and one person acts as scribe by writing the understandings onto Post-its. Another person takes online notes to elaborate the words on the Post-its. The compiled notes represent meeting minutes. The team review is done by reading aloud and stopping when we want to record something, or by reading a section silently and then discussing and recording the understandings.

We do not recommend streamlining the process until the first transcript is walked through in this way. This could take 4 to 8 hours for a 3-hour interview. By this time members of the team learn each other's focus and the team's focus has broadened enough to challenge their entering understandings of the system and users' work. Analyzing the first transcript is slow because it draws in related information from other transcripts and because the team is learning to work together. To streamline the process, people can identify key aspects of the transcripts before the meeting and then share and discuss them, screening redundancies. Also, individuals can each take one transcript (not their own), identify key aspects, and guide the walk-through of the transcript.

We record:

- A description of users' work
- The flow or structure of the work
- A description of problems in their work
- A description of problems with the computer tools
- Design ideas that emerge from our understanding of their work
- Questions for subsequent interviews

---

[3]Post-it is a registered trademark of 3M.

Questions naturally fall out of the inquiry process. Sometimes the questions direct us to specific users to interview. They can also indicate things that we need to find out from a particular person, or questions about technological possibilities.

A key to successful recording is to avoid structuring the information during interpretation. If we make duplicate points, they will be handled in the structuring process that follows. We also avoid judging the value of ideas. Decision making is separate from information analysis.

Structuring the Understanding

Some analyses have generated hundreds of Post-its. To work effectively with such a large amount of information requires some structuring.

Structuring contextual inquiry information is inductive not deductive: It is a bottom-up process. We use Affinity diagramming, a technique that supports inductive thinking. It is a way to synthesize qualitative information into conceptual groupings.

To begin, we divide the Post-its among the team members. The objective is to systematically group Post-its that are conceptually related while avoiding the use of predetermined categories. We want the groupings to emerge from the information.

Developing an Affinity Diagram is a team process that requires facilitation by one of the team members. We begin by selecting an arbitrary piece of information, placing it on a white board and asking: "What relates to this piece of information?" Someone else puts a related piece of information on the board and the discussion begins. As information is put on the board, people talk about the relationships they see or do not see between the information.

As the groupings emerge, individual items move between groups. This process continues until the groups seem fairly stable. When the groups stabilize, the team systematically reviews each group and assigns the group a name. The name might come from an individual item in that group. The group names are intended to crystallize our understanding of the specific items in the group. It is through the conversations about the items that a shared meaning is developed within the team. The meaning of the information guides the design and development effort.

When the groupings are stable, the team reviews and labels each group. We suggest limiting the individual items in each grouping to five. This forces new concepts to emerge so that the detail and complexity of the information is revealed. The first level groupings can then be further grouped into a second and third level grouping to form a hierarchy. Figure 9.2 shows the hierarchical structure of an Affinity diagram.

When complete, the headings represent domains of work and design areas for the system. The individual items represent design ideas and the details of users'

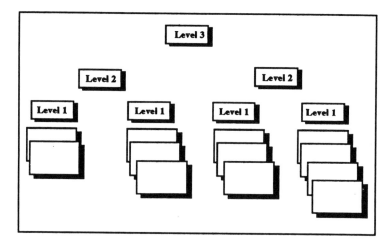

FIG. 9.2. Affinity diagram.

work. The team then reviews the "wall" to determine what is missing in the items. These missing items are recorded and direct the next focus.

We recommend working with no more than 300 Post-its at a time. Multiple Affinity diagrams can be completed as more information is collected. We do not recommend deliberately trying to integrate later Post-its into the first affinity headings. Rather a new structure should be created. When these structures are placed side by side, they represent a full picture of the information. Overlapping headings can be integrated later.

We have found it effective to use large pieces of paper as a background for the Affinity diagram. Flipchart paper works well. We tape the paper vertically so that it can be rolled up like a scroll, removed from the room, and reused as necessary. It is also effective to use white boards as the Post-its surface, and then place the groupings in a notebook for portability.

When the Affinity diagram is recorded, it represents an effective means of communicating the emerging system vision. This information can be used in requirements documents, decision-making processes, and to focus design meetings.

USING CONTEXTUAL INQUIRY WITH OTHER
PARTICIPATORY DESIGN TECHNIQUES

Contextual inquiry is a method that stimulates user's awareness of their work practices. By being in the context of the user's work, the user is surrounded by the artifacts of work. While doing their work they simultaneously reflect upon their work. Thus, in-context interviewing fosters awareness of work practice. Through partnership we engage users in reflection on their work experience and help them articulate the nature of the work. Through focus we direct the dialogue and develop a model of user work. This model is a description of the user's current work practice. The work model can be represented in a graphical or textual form.

Implicit in every software system is a model of how users work. Contextual inquiry provides a way to ground the system work model. The model of user work that we derive through contextual inquiry becomes the basis for the system model, and subsequently the user interface representation. By combining contextual inquiry with other participatory design techniques, such as paper prototyping, we engage users in the design of a user interface that supports the work model.

Contextual inquiry has been used effectively throughout the development cycle to ground design action in the user's work experience. Other participatory design techniques, such as Metaphor Workshops, also serve the purpose of stimulating awareness and providing a structure for involving users in design.

We have conducted Metaphor Workshops with users as part of our design team meetings (Kensing & Madsen, 1991). In Metaphor Workshops, metaphors are used to stimulate awareness of work. Everyday metaphors are used to help users think about their work. Metaphor Workshops use metaphors as the stimulant instead of the actual work environment. The workshop involves a design phase where design ideas are generated with users. The design ideas tend to be discrete ideas, rather than a coherent work model. Similarly Future Workshops, which we have used less frequently, have the same elements of a stimulation technique and a design phase.

We have brought users into engineering design meetings to develop work scenarios. The focus on scenario building stimulates an awareness of user's work practices. The questions and experiences of the other meeting members serves to increase awareness. An outcome of these meetings is a model of user work. Through design meetings with users and engineers, the work model is transformed into a system design.

We have made extensive use of paper prototyping (Ehn & Kyng, 1991) in conjunction with contextual inquiry at customer sites and at internal design meetings. Contextual inquiry techniques stimulate awareness of work practice; paper prototypes are used to model work practice and user interface designs. When paper prototypes are used during the interview, or when they are brought back to the user following generation in design meetings, the user can see what has been said and respond to the interpretation that is now represented physically. The paper prototype provides a way for users to alter the work model or user interface design directly.

We use Post-its to model the work and user interface design. Paper prototypes of the user interface using different size Post-its for windows, menus, and dialog boxes provide a representation of the system that users can easily question and manipulate.

Paper prototypes are advantageous because they can be made quickly and support a continuous challenge and change model of design. They support fast iteration and an inquiry process that challenges assumptions and ideas. When generated by design teams, paper prototypes also allow for multiple versions of a user interface to be mapped out for codesign with users. Figure 9.3 presents an example of paper prototyping a user interface.

After the prototypes are created, we take these to users and use contextual inquiry to get their reaction. This "what if" conversation allows users to co-design the system as they define what they think functions should do and ask for additional features. Together, designers and users engage in a conversation about the design.

Participatory design practices are techniques that stimulate user's awareness of their work so that work practices can be articulated and users can participate in the design of a system that will support their work. Contextual inquiry is a

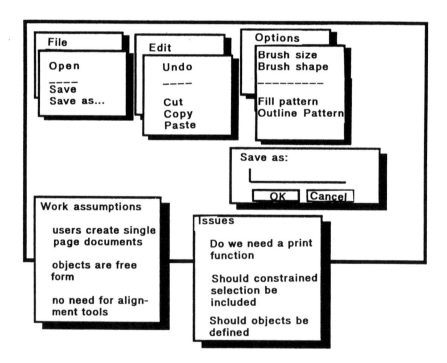

FIG. 9.3.   Paper prototype.

technique that achieves this end. When used in conjunction with other participatory design techniques we gain the benefits of multiple forms of user stimulation and ways to generate design ideas. A result is that users feel that we are incorporating their system needs and invite us to come back.

USING CONTEXTUAL INQUIRY THROUGHOUT THE SYSTEM DEVELOPMENT CYCLE

Each phase of development has key questions and tasks to which engineering teams must respond. Although our design practices are generally iterative and involve simultaneous requirements definition and design, we use a water-fall model of development to show where and how contextual inquiry can be used throughout the development cycle.

The question during the first phase of development is: "What should we build?" The key tasks are to determine a product strategy and product requirements.

We have used contextual inquiry to gather system requirements for a variety of products and systems. We use contextual inquiry with users of our systems and competitive systems. The focus for these inquiries is:

- What is the user's work?
- What tools are currently used?
- What works well and why?
- What are the problems that we can address with our technology?

Inquiry with users results in an understanding of system requirements that becomes the first cut at a product requirements document. Information at this stage is also used with planning tools, such as Quality Function Deployment (QFD). We use QFD to map user's needs to system requirements (Cohen, 1988). It provides a systematic process that is sensitive to customers' business needs, engineering resources, and marketing needs. QFD is one way that we separate design from decision making.

The requirements gathering phase is an opportune time to establish partnerships with strategic customers that we can work with intermittently throughout the development life cycle.

The second phase of development is focused on project planning and preliminary design. We have used contextual inquiry at this stage with paper prototyping to define the system work model. This is also the time that we use Metaphor Workshops, scenario building, and user participation in design meetings.

Design and coding of the implementation are the key tasks of the third phase of development. We have used contextual inquiry to codesign the system work model and user interface with users. Initially, the designs are represented as paper prototypes and later as software prototypes or early versions of the software system. When software is available for restricted use, we conduct *test drives*. Test drives provide users with an opportunity to explore the system while engaging in dialogue with designers about work flow, work concepts, and interface mechanisms. Although some test drives do not take place at the user's work site, designers use the principles of contextual inquiry in their interaction with users.

External field test of the system occurs during the fourth phase of development. During field test and after the product ships, we conduct contextual inquiry sessions with users in their environment while they are using the system for their work. Initially, we focus on ease of learning and later on experienced usage of the system.

Throughout the development cycle, user collaboration and system design are part of an ongoing, iterative process. The results of the inquiries are incorporated into the design of the system. After the product ships, results of inquiries can be incorporated into subsequent versions of the product or into a new product. The system development cycle in Fig. 9.4 illustrates how contextual inquiry can be used throughout the development cycle.

The type and extent of impact that contextual inquiry has on a system depends on when it is used. At the beginning of the product cycle, there is greater opportunity to define the product. This is the time to determine fundamental issues related to product functions and system structure that supports user's work flow and work concepts. Later in the development cycle, the design focus narrows to issues specific to the user interface, such as interface terminology and sequence of steps. The design of a system work model that supports user work and the design of the user interface is equally important. Working with users throughout the development cycle ensures that we consider both of these areas, not simply user interface issues.

Contextual inquiry is used as a method for collaborating with users throughout the system development cycle. As such, it provides a way for users to participate in the design process. This participation can be limited to a contextual inquiry session that focuses on user's work and tool use, or it can be expanded to include codesign of the system through the use of prototypes and participation in design meetings. In Digital, we are combining contextual inquiry with other participatory design practices and quality decision-making processes. When used together we have been able to powerfully impact the direction of products and the ways that design teams work.

We invite you to try contextual inquiry. Take from us what we have learned, try it in your organization, and tell us how you have improved upon our ways of bringing the users' voice into the design and development process.

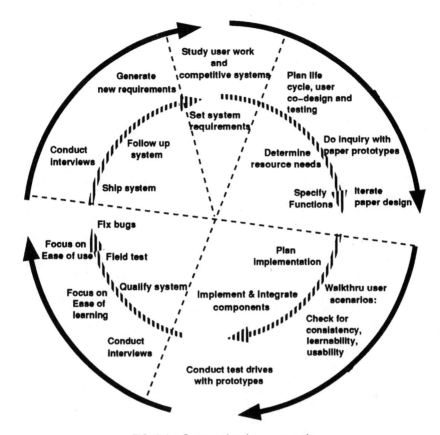

FIG. 9.4.   System development cycle.

ACKNOWLEDGMENTS

The efforts of several people contributed to this chapter. We especially thank Melinda Thomas for her careful reviews of this chapter. We also thank Hugh Beyer, Michael Catano, Julie Dingee, Anne Duncan, Michael Good, Glen Lupton, Claudia Marcus, and Dennis Wixon for their comments. We thank John Bennett, Steve Knox, and Dennis Wixon for their help in developing the contextual inquiry course on which this chapter is based.

REFERENCES

Bennett, J., Holtzblatt, K., Jones, S., & Wixon, D. (1990, April). *Usability engineering: Using contextual inquiry.* Tutorial presented at CHI '90: Conference on Human Factors in Computer Systems, Seattle, WA.

Blomberg, J., & Henderson, A. (1990). Reflections on participatory design: Lessons from the Trillium experience. In *Proceedings of CHI'90: Conference on Human Factors in Computing Systems* (pp. 353–359). Seattle, WA. New York: Association for Computing Machinery.

Cohen, L. (1988). Quality function deployment: An application perspective from digital equipment corporation. *National Productivity Review, 7*(3), 197–208.

Darroch, V., & Silvers, R. (Eds.). (1982). *Interpretive human studies: An introduction to phenomenological research.* Washington, DC: University Press of America.

Ehn, P. (1988). *Work-oriented design of computer artifacts.* Stockholm: Arbetlivscentrum.

Ehn, P., & Kyng, M. (1991). Cardboard computers: Mocking-it-up or hands on the future. In J. Greenbaum & M. Kyng (Eds.), *Design at work: Cooperative design of computer systems* (pp. 169–196). Hillsdale, NJ: Lawrence Erlbaum Associates.

Gentner, D., & Grudin, J. (1990). Why good engineers (sometimes) create bad interfaces. In *Proceedings of CHI'90: Conference on Human Factors in Computing Systems* (pp. 277–282). Seattle, WA. New York: Association for Computing Machinery.

Glaser, B., & Strauss, A. (1967). *The discovery of grounded theory: Strategies for qualitative research.* Chicago: Aldine.

Good, M. (1989). Seven experiences with contextual field research. *SIG-CHI Bulletin, 20*(4), 25–33.

Gould, J. D., Boies, S. J., & Lewis, C. (1991). Making usable, useful and productivity enhancing computer applications. *Communications of the Association for Computing Machinery.*

Grudin, J. (1990). Obstacles to participatory design in large product development organizations. In *Proceedings of PDC'90: Participatory Design Conference* (pp. 14–21). Seattle, WA. Palo Alto: Computer Professionals for Social Responsibility.

Heidegger, M. (1962). *Being and time,* (J. Macquarrie & E. Robinson, Trans.) New York: Harper & Row.

Holtzblatt, K., Jones, S., & Good, M. (1988). Articulating the experience of transparency: An example of field research techniques. *SIGCHI Bulletin, 20*(2), 45–47.

Jones, S. (1989). *Learning DECwrite in the workplace: Using contextual inquiry to articulate learning* (Digital Equipment Corp. Tech. Rep., DEC-TR 677). Nashua, NH: Digital Equipment Corp.

Kensing, F., & Madsen, K. (1991). Generating visions: Future workshops and metaphorical design. In J. Greenbaum & M. Kyng (Eds.), *Design at work: Cooperative design of computer systems.* Hillsdale, NJ: Lawrence Erlbaum Associates.

Kuhn, T. S. (1962). *The structure of scientific revolutions.* Chicago: University of Chicago Press.

Kyng, M. (1988). Designing for a dollar a day. In *Proceedings of CSCW'88: Conference of Computer-Supported Cooperative Work* (pp. 178–188). Portland, OR. New York: Association for Computing Machinery.

Nygaard, K. (1990). Program development as a social activity. In *Proceedings of PDC'90: Participatory Design Conference* (pp. 4–13). Seattle, WA. Palo Alto: Computer Professionals for Social Responsibility.

Packer, M. J. (1985). Hermeneutic inquiry in the study of human conduct. *American Psychologist,* pp. 1081–1093.

Whiteside, J., Bennett, J., & Holtzblatt, K. (1988). Usability engineering: Our experience and evolution. In M. Helander (Ed.), *Handbook of human-computer interaction* (pp. 791–817). New York: North-Holland.

Winograd, T., & Flores, F. (1986). *Understanding computers and cognition.* Norwood, NJ: Ablex.

Wixon, D., Holtzblatt, K., & Knox, S. (1990). Contextual design: An emergent view of system design. In *Proceedings of CHI'90: Conference on Human Factors in Computing Systems* (pp. 329–336). Seattle, WA. New York: Association for Computing Machinery.

10 PICTIVE: Democratizing the Dynamics of the Design Session[1]

Michael Muller*
Bellcore, Piscataway, NJ

This chapter describes a year's practice involving what appears to be a promising new system design technique: PICTIVE, or Plastic Interface for Collaborative Technology Initiatives through Video Exploration. The technique employs a combination of deliberately low-tech design components with high-tech video recording facilities.

The "P" in PICTIVE stands for "Plastic," and is meant to do so in at least three ways:

- *Composition.* Many components in the design technique are literally made of colored plastic. The relative durability and inexpensiveness of the plastic parts appear to encourage the participants to be bold with them, and this in turn stimulates inventive design and creative problem resolution.

- *Ease of Change.* This leads to the second sense of the word "plastic:" the interface objects and concepts are seen as being very malleable, and the participants can try many variations quickly and easily.

- *Artificiality.* The third intended connotation of *plastic* is artificiality: Unlike other prototyping paradigms, the PICTIVE technique cannot be confused with a working system—or even with technical feasibility. With a software-implemented prototype, it is sometimes difficult to explain why a working prototype does not translate quickly into a working, production-grade system. With the PICTIVE techniques, there is no software prototype (although we hope that

---

[1]This chapter is an expansion and combination of four papers (Muller, 1991a, 1991c, 1991d, 1992), and is published here with permission of the Association of Computing Machinery.
*The author is currently at US West Advanced Technologies, Boulder, CO.

211

the design clarification and understanding that take place during PICTIVE can lead to a software prototype). As a result, users' and developers' expectations of one another may be more easily managed after the design has been agreed upon.

I begin with a review of the background rationale for the technique and the design paradigm that it supports. This is based on two research and practice traditions:

- the past two decades' experiences as reported by the Scandinavian school of industrial design founded by Kristen Nygaard (e.g., Bjerknes, Ehn, & Kyng, 1987).
- visual communication paradigms, as explored and clarified by Bly and colleagues (Bly, 1988; Bly & Minneman, 1990; Harrison & Minneman, 1990; Minneman & Bly, 1990; Tang & Minneman, 1990), by Crane (1990, 1991), and by Verplank (1990).

I then describe the PICTIVE technique as we currently understand it, followed by a comparison between PICTIVE and the Scandinavian techniques from which it arose, as well as related techniques in use in Britain and North America. I continue with a description of the role of PICTIVE in several products and projects, as well as its potential uses in other projects. The subsequent section provides an initial assessment of the technique, including a conceptual analysis of why the PICTIVE technique works as well as it seems to. I close with a brief analysis of how participatory design may fit into our North American workplace.

PARTICIPATORY DESIGN

Designing With the User

The "Scandinavian Challenge" (e.g., Bjerknes, Ehn, & Kyng, 1987) or participatory design school emphasizes a tradition of user participation in workplace decisions in improving the quality, productivity, and satisfaction related to computing systems (Ehn, 1988a; see also Suchman, 1988a, 1988b; for a related argument, see Grudin, 1988). An important aspect of this approach is that users act as fully empowered participants in the design process (Bodker, Ehn, Kammersgaard, Kyng, & Sundblad, 1987; Bodker, Ehn, Knudsen, Kyng, & Halskov Madsen, 1988; Bodker, Gronbaek, & Halskov Madsen, 1990; Ehn, 1988a, 1988b, 1990a, 1990b; Ehn & Kyng, 1987; Floyd, 1981, 1987; Thoresen, 1990a, 1900b). User participation provides the opportunity to affect decisions about systems that will later impact users' work lives. One technique for doing this is to insure that users have early exposure to the target implementation technology—even if this must be done through mock-ups (Bodker et al., 1987, 1988). In this way, the users can understand those aspects of the technology that directly impinge on their jobs (e.g., Kraut, Dumais, & Koch, 1989) and their lives (Bruce & Kirup, 1987; Darwin, Fitter, Fryer, & Smith, 1987). Users can then bring this improved understanding to bear on negotiations about what attributes and features will actually be implemented (Ehn & Kyng, 1987).

Critique of Software Rapid Prototyping

The PICTIVE technique provides a different kind of mock-up—a paper-and-pencil concretization of what the system will eventually look like and how it will behave. In this way, PICTIVE falls within the tradition of participatory design via the provisions of mock-ups (e.g., Bodker et al., 1987, 1988).[2] But while PICTIVE attempts to address the broad issues of user participation in the design *team*, it is also concerned with a local or immediate openness of the design *session* to effective user participation. PICTIVE is intended to facilitate a partnership model (e.g., Reason & Rowan, 1981) in which developers and users work together to design a software system.

The issue for PICTIVE is the technology that is used in rapid prototyping paradigms. Typically, this technology has taken the form of a sophisticated software system. The user's ideas are expressed within the software prototyping system, usually through the intermediary services of a human expert. The expert is typically a developer or an implementor, and so there is an inherent power imbalance: The "owner" of the prototyping technology (the expert) has control over the rapid prototyping process, and the user's ideas must always be interpreted by the owner, because the user lacks the training to operate the software prototyping system.

In this analysis, software rapid prototyping technology contains implicit politics (e.g., Winner, 1980; see also Suchman, 1988b) that have implications for the relationship between the user and developers, implementors, or designers. If the user lacks the time or inclination to be trained in the prototyping software, then the user is dependent upon a software professional whose personal or organizational agenda may be quite different from the user's. As a result, the user's ideas may be distorted, and the social or organizational effort required from the user to correct the distortion may be prohibitive. The user may thus be alienated from the design process and from the artifact produced by the process.

Alternatively, users who *are* trained in the prototyping technology are at risk of alienation from other users, as their prototyping-environment-influenced views become associated with those of the designers and implementors, and not

---

[2]It should be emphasized PICTIVE is one approach to one aspect of the realm of problems and approaches addressed via participatory design. For others, see Bjerknes et al. (1987), Floyd et al. (1989), Greenbaum and Kyng (1991), Muller (1991b), and Namioka and Schuler (1990).

with those of their original coworkers.[3] In this scenario, the users who are *not present* in the design process may not be faithfully represented by the now-trained designing user, and may be alienated from the resulting artifact (Bodker et al., 1987).

One of the motivations of PICTIVE is to use a low(er)-tech prototyping technology—paper-and-pencil, markers, plastic "icons," Post-It™* notes, and the like—to insure "equal opportunity" access to the prototyping technology by all participants in the design session.[4] Because PICTIVE requires minimal training, we hope that it reduces the dangers of alienation. Simultaneously, PICTIVE uses relatively high-tech recording techniques to preserve a detailed record of the design session.

Design Records, Documentation, and Communication

Our experience touches on a second important area: the communication of designs to implementors. In large North American organizations, the software development life cycle tends to be a multiple-person, multiple-organization process with numerous opportunities for miscommunication (e.g., Grudin, 1990a, 1990b). Requirements documents are notoriously unsuccessful in providing useful information to guide the implementors' design decisions. We take the approach of providing what might be the video equivalent of a design dairy (e.g., Naur, 1983) which may be communicated from the design session participants to the implementor(s). The video record of the design session serves both as a dynamic presentation of the design, and as a conversational rationale for that design and for the decisions underlying the design.

Participatory Research

This chapter describes both a design paradigm and an informal research strategy for studying that paradigm. The underlying research foundation is contextualist (e.g., Whiteside & Wixon, 1988), emphasizing an engagement or involvement of the research with the "researchees," rather than the more classical role of researcher as observer, and research "subject"—i.e., the user—as object (see also Reason & Rowan, 1981). This reduces the objectification of the participants (Thoresen, 1990b), and maintains all participants' roles as active determiners of

---

[3]In a recent usability study, user representatives on three-year rotations to the development organization stated their full satisfaction with a user interface design that, *in the field,* proved very difficult to use (unpublished data of Muller, Hornstein, Flamm, Ewedemi, Gebele, and Sauer; see also Bush, 1990).

*Post-It is a trademark of 3M Corporation.

[4]The disability status of some participants may reduce the "equal access" nature of the technique. This problem is pervasive in the workplace, and requires a more general solution than has been attempted in this project.

the system's attributes. By contrast, more conservative "acceptance testing" techniques tend to treat the user as a measurement instrument (a different class of object) that takes a subordinate, quality-assurance role with respect to the traditional focus of attention, the software. Floyd (1981, 1987) analyzed these two approaches, contrasting the more user-centered "process-oriented paradigm" with the more conservative "product-oriented paradigm."

PICTIVE in Context

The PICTIVE technique may appear at several points within the software development life cycle context. The technique may be used for early requirements gathering and analysis—a kind of egalitarian knowledge acquisition. It may be used for design work—or for redesign of a system that is already in the field. It may also serve a conflict resolution role, by facilitating concrete, clear discussions of certain aspects of system functionality that have generated conflict or disagreement.

PICTIVE may also be used iteratively, in a streamlined version of software rapid prototyping. Or PICTIVE sessions might be alternated with software rapid prototyping episodes, to use two different but related techniques for improved communication among the codesigners. This might be done in an initial-development context, or in an iterative approach to usability testing.

PICTIVE METHOD

The general notion of PICTIVE is that the design team manipulates concrete objects to explore a number of scenarios about how the target system is supposed to support the user's job or task. Meanwhile, video recording equipment captures their session for later playback, analysis, issue exploration, and eventual use as a video form of a design document, to be used to guide implementation, and to explain the rationale behind certain design decisions.

Apparatus

The apparatus for PICTIVE includes video recording technology and a collection of design objects that are manipulated on a design surface (Fig. 10.1). The video equipment is used to record a design session. While we have explored some elaborate split-screen techniques using a video special effects generator, the simplest procedure is to focus the camera on the design objects and to record the voices of the design team as they manipulate those objects. It is helpful if one member of the team can view a monitor to be certain that the intended configuration of the objects is being recorded. So far, there appears to be no requirement that the camera be hidden or even unobtrusive (Fig. 10.2); however, we will have

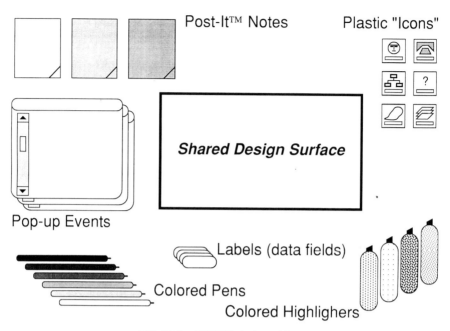

FIG. 10.1.   PICTIVE design objects.

to reexamine this claim as we use the technique with a greater variety of users (see the section on Assessment).

The design objects mentioned earlier fall into two categories. The first and simplest category is every-day office materials. These include: colored pens, colored high-lighters, colored papers, colored Post-It notes of various sizes, colored stickers and labels, and colored paper clips. A more detailed list, with rationales for some of the items, appears in Appendix 1.

The second category is materials prepared by the developer—either *generically* for multiple design exercises (e.g., command line, query fields, menu bars, dialogue boxes, etc.), or *specifically* for the project being designed. For PICTIVE-based design of an experimental project management groupware prototype (Muller, Smith, Goldberg, & Shoher, 1991), for example, we used a set of plastic squares in assorted colors with icon designs drawn on them (Fig. 10.3), a suite of paper images of pop-up events (Fig. 10.4), and several sheets of desk-blotter paper that corresponded to large, fixed areas of a presentation surface (Fig. 10.2). For a user-driven redesign of a computer-generated text printed report, we created modifiable components of that report on large sheets of easel-pad paper, and made alternative versions of subsections of the report on colored construction paper (Muller & Taverner, unpublished data). Details of the construction of these items appear in Appendix 2.

FIG. 10.2.   Conducting a PICTIVE session: An "equal opportunity" design surface with video recording.

In addition, several tools are provided with which to modify these items, including brightly colored scissors, erasers, and the colored pens and high-lighters mentioned before (for a rationale for the use of color, see later section on "Why the PICTIVE Technique Seems to Work.").

PICTIVE "Homework Assignments"

A PICTIVE session would not be very productive unless the participants had given their shared work some thought beforehand. Therefore, each participant is asked to prepare a homework assignment. Typically, these have been expressed as follows:

- *Users: Job/Task Scenarios.* "Think through what you would like the system to do for you. Be concrete and specific. Be prepared to go through the steps that would be required to 'do the job' (or 'do the task'). Remember: you are 'The Expert' on the content of the job, and you will contribute your unique knowledge and expertise to the cooperative design session. But don't be surprised if your ideas change during the session."

- *Developers: System Components.* "Based on prior discussions with the user, construct an initial or preliminary set of 'system components' that the users

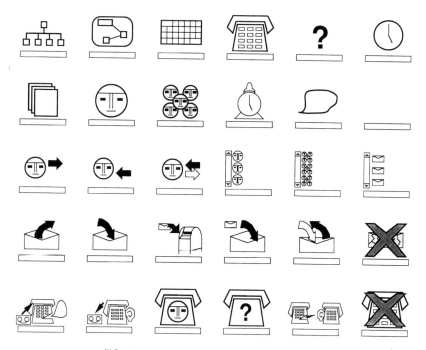

FIG. 10.3.   A sample set of PICTIVE plastic icons.

will manipulate (and change!) while exploring the job/task scenario. Be concrete and specific. Remember: you are 'The Expert' on the system and its environment, and you will contribute your unique knowledge and expertise to the cooperative design session. But don't be surprised if your preliminary view of the 'system components' changes during the session."

In the expanded version of the procedure that was used in the *Practicum in Methods for User Centered Design* (White, Wildman, & Muller, 1991), there were six roles: user, developer, human factors specialist, market specialist, tester, and technical writer. Each role had its own, domain-relevant homework assignment. Some of the user roles may require greater expansion—for example, to include multiple categories of users, such as people who use printed reports, and support and installation staff (Anderson, 1990). This theme is pursued in Muller (1991a), and in the section called "Who Participates in Participatory Design," later.

PICTIVE Process

In general, the PICTIVE session proceeds through the mutual education in one anothers' perspectives that Bodker et al. (1990) advocate. Each participant brings her or his expertise, based on job function and background, to bear on the

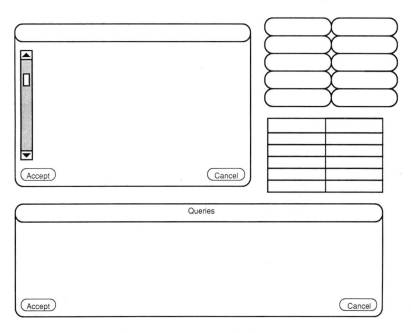

FIG. 10.4.   A sample set of PICTIVE pop-up events.

group's shared design problem. If successful, the resulting design is not *driven* by any single participant, but represents a synthesis of the different participants' different views. So far, we have not seen a need for a designated leader or coordinator; however, we anticipate that it may be necessary to designate one of the participants to watch an agenda (generally agreed-upon through consensus) and keep the group on track.

In addition to the physical model of PICTIVE (low-tech objects, etc.), there is also an emerging process model.[5] Ideally, each participant carries out a homework assignment prior to the session, and the session begins with a review and introduction to the homework assignments that have been prepared by the various participants. The users' work scenarios are then explored in detail, leading to a discussion of how the developers' technology can be applied to meet the users' human needs. Other participants introduce their needs into the discussion (e.g., designing for testability, designing for marketability, etc.).

The attributes of the emerging process model are as follows:

• *Reciprocal Education.*   Each participant has the responsibility to explain her or his perspective, personal stakes, and competence to the other participants.

[5]I thank Elizabeth Dykstra Erickson for pointing out the importance of a process model as a component of PICTIVE.

• *Reciprocal Preparation.* These explanations are most easily done if each participant prepares for the session—for example, the developer might bring a working, general example of a new technology, the human factors worker might bring paper-and-pencil components of the system, and the user might bring step-by-step scenarios of use.

• *Reciprocal Validation.* The purpose of the session is not for all participants to be convinced of one world view or one vocabulary, but for each participant to understand and give credence to the points of view and vocabularies of the other participants (e.g., Blomberg, 1990, 1991; Bodker et al. 1990; Bonavoglia, 1991; Suchman & Wynn, 1984; see also works on applied constructivism, e.g., Belenky, Clinchy, Goldberger, & Tarule, 1986; Guba & Lincoln, 1989; Schon, 1983). This permits the participants to collaborate across disciplinary and organizational boundaries through mutual understanding, while nonetheless remaining faithful to their own experiences, personal stakes, and workplace realities, as well as to the other people whom they represent.

• *Checking In.* Because the purpose of the session is to hear the views and perspectives of all the people in the group, the participants usually develop a pattern of making sure that no one is being left out or ignored.

• *Emergent Designs.* Often, the design becomes something that none of the participants initially envisioned, through the combination of expertises and perspectives that were, as a set, unavailable to any single participant.

• *Consensus Decision-Making.* The emergence of new design ideas that contain multiple contributions helps to move the group away from entrenched positions. Often, decisions are made by consensus rather than by voting or other approaches that might disenfranchise the minority.

In most PICTIVE sessions, ideas expressed on the design surface are captured on the video recording, for use in interpretation and implementation. Although we initially feared that video recording would have an inhibitory effect, we discovered exactly the opposite. Because a video record is kept, there is no need for one participant to take the role of a formal record-keeper. This permits the participants to work together, without imposing a social distance between "designers" and "recorder," or between "subjects" and "researcher."

PICTIVE IN USE

This section reviews selected projects on which we have used PICTIVE, and what we have learned from them.

Project Management Groupware

The Project Management Groupware (PMG) project was our first experiment with participatory design. The goal was to build an experimental research prototype using an experimental AI software environment (Cebulka, 1990; Smith &

Cebulka, 1990), to support the project planning and project management needs of knowledge workers who were independently writing interrelated and interdependent documents. Our initial design experiences were highly positive: the users were enthusiastic about using paper-and-pencil techniques to express their needs and ideas, and the human factors workers believed that they had found a new means for obtaining requirements information. We reported our design as a success story (Muller, Smith, Goldberg, & Shoher, 1991).

Later, we discovered several limitations of what we had done. The most important problem was the poor transfer of the system concept from the design sessions to the developer of the research prototype. This occurred because the developer was not involved in the participatory design sessions. Our initial hope—that the video records would serve as a simple video design document— also turned out to be naive: The design sessions produced over a dozen hours of videotape, and the developer simply didn't have the time to comb through those records to collect information for implementation.[6]

A second weakness—and this was compounded by the absence of the developer—was our initially poor understanding of a process model for this type of participatory design activity. The PICTIVE sessions tended to approve any suggestion that the users made. This turned out to present significant problems to the developer, whose experimental environment did not have the functionality to support certain design options preferred by the users. If the developer had been part of the PICTIVE session, she or he could have initiated a discussion of design alternatives that were more technologically feasible, and perhaps could have led users to new technology-based capabilities that might have enhanced the workplace value of the target system.

The result was a prototype that functioned over a portion of its intended domain. However, it became clear that further work could not be based on either the technology in this prototype or the design from the initial PICTIVE sessions: The gap between intended design and technology capabilities was simply too great. Thus, the absence of the developer from the PICTIVE sessions ultimately weakened the generalizability and the growth potential of the resulting prototype.

Document Design

An important user interface to one product involved computer-generated printed reports when a distributed computing process had failed. The reports took the form of notifications to craft personnel about what had gone wrong, and instruc-

---

[6]Reviewers of PICTIVE video records have commented that it is sometimes difficult to understand who is talking about what, because there are sometimes many voices and occasionally two parallel conversation. This is especially difficult if the reviewer was not present during the recording session. We speculate that stereophonic recordings of the audio portion of the video may provide spatial cues to help in this task, but we have not yet tried the experiment.

tions about how to fix it. One feature of the report proved unexpectedly complex and difficult to use. A PICTIVE session was held with eight management representatives of user organizations from the Bell Operating Companies to make the printed reports more job-oriented (Muller & Taverner, unpublished data).[7]

This exercise showed the value of Crane's (1990) advice, "When in doubt, DRAW!" and helped to confirm the importance of the highly visual and concrete nature of the PICTIVE materials. Initially, the representatives split into two groups with opposing plans for the refinement of the printed report. One group's position was taken on the basis of highly abstract arguments, which were difficult for the other group to contest. When the first group was persuaded to make their arguments visible and concrete—that is, to design their idea on paper—both groups rapidly understood the flaws in the abstract argument, and could then agree upon a modification of the second group's design. The resulting design was a consensus of the views of the different representatives, and was well-received by users and developers.

Graphical User Interface Redesign of a Text-Based Interface

In a third project, we made fewer and different errors. This was a graphical redesign of a text-based system. The PICTIVE sessions for this project were actually conducted by the developer, thus obviating some of the most salient mistakes of the PMG project. A more diverse group of participants was involved, including the developer, a user, a graphical analyst, a marketer, and a human factors worker.

The design effort was a success: All the participants believed that they had had the opportunity to express themselves, and that they had collaboratively solved their problems. The resulting research prototype in fact worked fairly well. However, the users were somewhat disappointed with certain attributes, leading to another design and implementation iteration.

The problem appeared to center on miscommunications regarding the implementation technology. Several members of the design team were accustomed to "dumb" terminal technology, whereas other member's mental models were the X-Windows™* system. The result was undetected miscommunication about certain of the dynamics of the target environment, and consequent mismatch of expectations about a few features of what was to be built. The lesson that we learned from this experience was that the developer's (or technologist's) role in PICTIVE is to provide a *detailed understanding* of the technology. This level of detail is difficult to achieve by paper-and-pencil techniques.

---

[7]Appendix 2 contains a discussion of the materials used for this printed-page user interface.
*X-Windows is a trademark of MIT.

Facilities Allocation System

The problem of technology explication appears to have been solved in work on a facilities allocation system product, called FLEXCOM™/linc.* Participatory design of a portion of FLEXCOM/linc (Muller, Kaye, & Sauer, 1991) was *preceded* by a design walk-through based on a partially implemented prototype; the walkthrough emphasized both user-oriented functionality and technology-oriented capabilities. Thus, by the time that the PICTIVE session began, the users had undergone a 45-minute tutorial on the target technology. These PICTIVE sessions usually involved two users, a developer, a marketer, and one or more human factors workers.

One notable aspect of these sessions was that the users and the marketers pressed the developers and the human factors workers to take full advantage of the sophisticated capabilities of the target windowing environment in order to meet detailed user-initiated design requests. This suggests that the design walkthrough had adequately communicated the developer's technology ideas, supporting a richer interaction to meet the users' needs. However, these provisional conclusions will have to be reexamined in late 1991, when the product is subjected to full usability testing.

Classroom and Workshop Use

PICTIVE has been used in two classroom settings. The *Practicum in Methods for User Centered Design* (White, Wildman, & Muller, 1991) teaches a number of methods, including PICTIVE as its representative case for participatory design. PICTIVE has also been used in a tutorial workshop on graphical user interface design with the twin goals of (a) teaching about the OSF/Motif™* Style Guide (Open Software Foundation, 1991), and (b) exploring the adequacy of the OSF/Motif Style Guide to meet certain classes of business needs. Both classroom experiences led to the same lesson about PICTIVE practice: There should be a clear decision about how fully specified the requirements are to be, *before* the PICTIVE session begins. With a detailed specification, PICTIVE can be used immediately to begin collaborative design. With a high-level specification, the session often becomes an extended requirements analysis instead, and little concrete design work is accomplished within the limited classroom schedule. Of course, requirements analysis work is valuable, too. The point is not to discourage concretized requirements analysis, but simply to insure that an experience that is oriented toward *design* should have a detailed set of requirements already in hand, to support that design work.

---

*FLEXCOM is a trademark of Bellcore.
*OSF/Motif is a trademark of Open Systems Foundation.

World of Intelligent Tutoring Systems (WITS)

One other problem surfaced when PICTIVE was applied to a research prototype in the World of Intelligent Tutoring Systems (WITS) project (an earlier report of this project appeared in Silverstein, Farrell, and Smith, 1990; see also Muller, Farrell, Cebulka, and Smith, 1992). PICTIVE was used in this context as a knowledge acquisition tool. Following fairly standard expert system paradigms, the researchers asked their domain expert to use PICTIVE in order to explain how WITS should execute the authoring and teaching of curricula, with a knowledge-based system substrate. The problem in this case was that the eventual end-users of the system—that is, the nonexpert authors and students—were not consulted. The system was in fact not a good match to their needs. This outcome is not surprising in view of the standard user centered design principle of *Know Thy User* (e.g., Helander, 1988; White, Wildman, & Muller, 1991). Nonetheless, it may be necessary to point out that PICTIVE is not particularly useful if the actual users of the system are not involved!

Selecting Projects for PICTIVE

In general, PICTIVE appears to be most useful in projects and products for which the requirements analysis has already been completed. Several practitioners have reported informally that, in the absence of clear requirements, PICTIVE sessions tend to remain abstract and unapplied; this has proven true in the course-based uses of PICTIVE, too.

PICTIVE and other hands-on techniques (e.g., Bodker et al., 1987, 1988; Dykstra & Carasik, 1991) are particularly well-suited for situations in which there is a need for *mutual* education—that is, of software professionals by the users, and of the users by the software professionals. Its strengths appear to lie in providing a straightforward bridging language (Dykstra, 1991a, 1991b) or shared linguistic code (Holmqvist & Halskov Madsen, 1989) that facilitates concrete communications across different domains of competence.

WHY THE PICTIVE TECHNIQUE SEEMS TO WORK

Assessments

We have tended to use PICTIVE in applied settings, in order to get a job done for a practical product. This has had an adverse effect upon formal assessments of the technique: We have attempted to obtain detailed questionnaire data from end-users and from developers on the use of PICTIVE. However, when we faced a choice between *doing the product work* and *collecting the data,* we have generally erred on the side of less data and more productivity. Therefore, our assessment is limited at present.

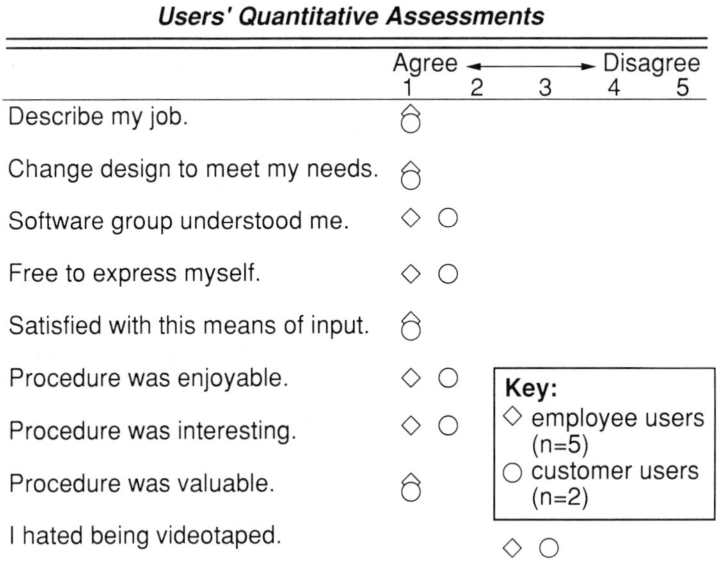

Users' Quantitative Assessments

	Agree ◄──────► Disagree				
	1	2	3	4	5
Describe my job.	⬦Ọ				
Change design to meet my needs.	⬦Ọ				
Software group understood me.	◇ ○				
Free to express myself.	◇ ○				
Satisfied with this means of input.	⬦Ọ				
Procedure was enjoyable.	◇ ○				
Procedure was interesting.	◇ ○				
Procedure was valuable.	⬦Ọ				
I hated being videotaped.		◇ ○			

Key:
◇ employee users (n=5)
○ customer users (n=2)

FIG. 10.5. Users' assessments of PICTIVE.

Formal Assessments

Figure 10.5 presents questionnaire results from two of the three sessions conducted in Muller, Kaye, and Sauer (1991).[8] Users (i.e., telephone company business customer end-users and staff end-users) completed a 9-item questionnaire that asked them to indicate the extent of their agreement with statements such as "The procedure helped me describe my job to the developers," or "The procedure helped me to change the design to meet my needs," etc. As Fig. 10.5 shows, users were quite pleased with the technique in this particular application.

The views of software development personnel have been assessed through a Bellcore-internal version of one of the courses mentioned above (White, Wildman, & Muller, 1991). Using a 5-point scale, participants rated the course in the highest possible category on attributes such as "enjoyable" and "valuable," and routinely listed PICTIVE as the most valuable and enjoyable component of the course, and as the component that should be given the most time in an expanded version of the course. Software engineering staff who have participated in PICTIVE sessions on their products echo these views.

---

[8]The work of the third session took more time than anticipated, and so the questionnaires were abandoned in that session.

Informal Assessments

In addition to the formal assessments, we have collected informal assessments in two ways: (a) remarks made by users during summing-up/postsession-review activities of various applications of the technique, and (b) open-ended question-naire items of the form "the best aspect of the procedure was ————" (collected in Muller, Kaye, and Sauer [1991] as well as in course work [an early version of White, Wildman, and Muller, 1991]).

Users' informal comments included the following:

- Why didn't we do this two years ago?
- [The best aspect of the procedure was] the freedom to change what was best for my job function.
- [The best aspect of the procedure was] being asked for input on something that is to be used, rather than being told.
- I felt that you were really listening to me.
- I'd like to do this again!
- Not enough time [in the mixture of design review and PICTIVE].

The informal assessment approach also allowed us to record the views of software professionals (developers, designs, and analysts). The following comments are representative of their responses:

- We found out things that we never would have discovered otherwise.
- Allowed for ideas to be shown more explicitly than conversation alone, with very little expense in terms of preparation time.
- That certainly opened my eyes!
- It was difficult to stay on the subject.
- Participatory design won't help you if you ask the wrong users!

Summary

Although these initial assessments are promising, we recognize the need for further assessments, and especially more formal ones. We will attempt to structure such methodological research into future practical applications of the technique.

Conceptual Analysis

There appear to be a number of attributes of PICTIVE that make it work well. Some of these have to do with the atmosphere that appears to surround a PICTIVE session. Other important attributes probably have to do with the democratizing influence of removing high-technology or high-abstraction tools from the design team's interactions with one another, at least during these design sessions.

Design Games

PICTIVE is deliberately informal and toy-like. Its apparatus contains many bright colors. Design team participants are encouraged to use as many colors as they like. The plastic nature of many of the tools also probably contributes to the toy-like quality of the components. The result appears to be an atmosphere of play, in which people draw pictures, cut out shapes, and design new components on an experimental, exploratory basis. PICTIVE falls into a class of design games used by the Scandinavian school, including word games (Ehn, 1988a, 1988b), and analogies (Bodker et al., 1990). The apparatus seems to encourage a mode of thought unlike analysis or engineering (e.g., Bodker et al., 1990), or even artistry (e.g., Crane, 1990).[9] People appear to have an enjoyable time with the apparatus and with each other.

We have recently extended this games-based approach to other exercises for group design of systems (Muller, Wildman, & White, 1992).

Shared Concrete Views of an Accessible, Common Problem Space

Unlike some other requirements-gathering techniques, PICTIVE may be used in a way that obviates abstractions. This tends to equalize the contributions of the disparate members of the design team (Thoresen, 1990a, 1990b), rather than emphasizing a specialized vocabulary that only "experts" know.

Because PICTIVE uses everyday objects in conventional ways, it is nonintimidating. Unlike some iterative design scenarios, there is no complicated prototyping technology. As a result, there are no issues of users having to explain their desires to an analyst (who "owns" the prototyping technology), or of users fearing embarrassment (through lack of ease with the technology), and therefore refusing to try out their ideas on the technology. Nearly anyone can use felt-tipped pens, can cut up a piece of paper, can rearrange a set of plastic squares to form a pattern. Users can work out their ideas directly, without needing interpretation or intervention by other members of the design team. Diverse perspectives brought by different team members (e.g., users, developers, technical writers, etc.) can be explored on a relatively equal basis.

The mock-up or *use model* approach of Bodker and colleagues (1987, 1988, 1989) is similar in nature; however, PICTIVE appears to go a step further. In the

---

[9]This may lead to problems—sometimes we need to apply an engineering analysis to design issues! However, because PICTIVE is only a part of the overall process of designing and developing a system, we hope that the necessary analytical thinking can be applied successfully at other stages of the process.

mock-up techniques, the developers help the users to envision the system by providing constructed concretizations of the technology. In these scenarios, the users seem to be placed in a relatively passive role, responding to technology that is defined by the developers, more or less as a *fait accompli*. By contrast, in PICTIVE the users and the developers redefine the technology collaboratively. This not only puts the users in a more active role, but also may shape the technology so that it better supports the users' needs and is more appropriately suited to the demands of the job. Preliminary results from the Project Management Groupware exploration support this notion: Collaborative redefinition of the technology by users and developers led to new and challenging (and welcome) implementation problems, as well as new social communications dynamics (see Muller et al., 1991).

Visual Communication

PICTIVE is strongly visual. It allows people to demonstrate what they want to see in the system, in a relatively clear and thorough way. People have high confidence that all the participants are seeing essentially the same thing, and that they understand one another's different perspectives (Bodker et al., 1990; Crane, 1990, 1991; Thoresen, 1990a) through that shared view. By analogy to groupware, this would be called WYSIWIS (What You See Is What I See—e.g., Stefik, Foster, Bobrow, Kahn, Lanning, and Suchman, 1987). The point, in PICTIVE, is to make sure that what we are all looking at is clearly, depictively, denotatively graphical, so that we believe in our communication process.

Concrete visualization has been discussed as a major facilitator of creative design work and intellectual work (e.g., McKim 1972; Nelms, 1957; Verplank, 1990), and as an essential tool for interpersonal communication about design (Bly, 1988; Bly & Minneman, 1990; Harrison & Minneman, 1990; Minneman & Bly, 1990; Tang & Minneman, 1990; see also Crane, 1990, 1991). This category of communication is poorly understood. It is not necessarily draftspersonship, but is rather informal and imprecise (Harrison & Minneman, 1990). However, it is not "art," either (Crane, 1990; Nelms, 1957; Verplank, 1990). We speculate that its informal and nonprofessional quality (i.e., nonprofessional in the sense that it does not involve *skilled* visual expression) is essential to its effectiveness—perhaps in ways analogous to the general importance of informal communication in organizational settings (Kraut, Egido, & Galegher, 1990). We look forward to further studies of why this mode of communication works so well.

Video Narrative

The use of video as the recording technology has several advantages. First, it tends to equalize participants' access to the recording technology, in contrast to the conventional researcher's private notes that are inaccessible to the people whose data are being recorded. Second, it provides a low-effort means of record-keeping. Third, it deemphasizes role or status differences between participants— that is, people feel less constrained as "experimenter," "subject," "observer," and so on. Fourth, it provides an extremely complete—yet informal—record of the design and of the decisions that led to the design.

WHO PARTICIPATES IN PARTICIPATORY DESIGN?

One of the early challenges that we faced was to find a place for participatory design in a North American corporate context. As noted by Grudin (1990a, 1990b), participatory design is more difficult to accomplish outside of the Scandinavian countries because of differences in legislative environment, workplace unionization, and scale and fragmentation of software development organizational models (see also Muller, 1991b).

We believe that participatory design has specific advantages in terms of relatively delimited and *product-oriented* (Floyd, 1987) business motivations: leveraging all the available expertise to improve product quality and market share (Bush, 1990; Muller & Cebulka, 1990), and improving commitment to the product or project, both in the customers and in the downstream software development staff. What is very interesting is that the same personnel or roles—for example, developers, users, human factors workers, marketers, technical writers, trainers, and testers—are involved in both motivations, and in a third, as well.

An analysis of the "leveraging expertise" argument leads to the inclusion of a diverse group of professionals. Quality products are based on diverse sources and categories of knowledge and expertise—of technology, of design principles, and of the users' actual jobs and needs. Quality products should also be designed for complete testability, and should be made in a way that makes them easy to explain in documentation or in training courses, and easy to present to potential customers at the point of sale. Thus, the expertise of developers, users, and human factors workers should potentially be augmented by the expertise of testers, trainers, technical writers, and marketers.

Similarly, products tend to be more successful if they receive the commitment of all the people involved in their implementation and use. Certainly the commitment of the customers to the product is desirable. Carter (1989) provides evidence that an increase in customer commitment through participation in design activities can lead to increased willingness of customers to try out new technology features. Thus, participatory design can be valuable not only in shaping the technology to the users' needs, but also in shaping the user to be willing to try the technology. In addition, the commitment of other professionals in the development process will help to make the product understandable to all the development staff as they work on it, and will help them to know how to work together on it, supporting models of group competence (Dykstra, 1991a) and shared success (Crane, 1991).

Finally, these same personnel or roles are suggested by the Scandinavian themes of workplace democracy through participatory design (Bjerknes et al., 1987). A glib summary of the workplace democracy theme in terms of United States history is *no mechanization without representation* (Muller, 1991a), which may be restated as follows: The people who will be affected by a software system should participate in decisions regarding that system (Bjerknes et al., 1987; Ehn, 1988a; Greenbaum & Kyng, 1991; Namioka & Schuler, 1990). The users receive the most obvious impact of the system; however, there are other workers who may also been understood as *stakeholders* in the system, because their careers are also affected: developers, technical writers, trainers, marketers, and testers. Thus, a broad interpretation of participatory design, based in United States software engineering practice, provides for the possibility that representatives of some or all of these personnel might be involved in design decisions. A commitment to nonexploitive approaches (e.g., Thoresen, 1990a, 1990b) requires their *direct* participation, rather than an *indirect* protocol through which requirements analysts attempt to serve as intermediaries.

While it may be more convenient in a corporate context to phrase our argument in terms of the first two motivations—expertise and commitment—this appeared to us to be unwise. As Winner notes (1980, 1986), a social or political motivation that is left unsaid may *become unsayable* and may thus be forced further into the closet. We have therefore continued to state the democratic nature of one of the motivations of the approach.

In this way, we support the inclusive nature of participatory design through three different but *convergent* motivations: product quality, product commitment, and democracy. Maintaining the convergence of these three motivations will be a key requirement for the continued usefulness of participatory design in our workplace.

CONCLUSION

Taken simply as a technique, PICTIVE appears to provide improvements in knowledge acquisition. However, the technique is intended to work within a context. Our context values user involvement because users are the experts on their own jobs—that is, because users possess an expertise that cannot be duplicated or replaced by personnel who are not job incumbents. By providing a means for the expression of that expertise, PICTIVE is intended to increase the effectiveness of users in causing software systems to be more responsive to users' needs and interests. Ehn (1990b) discusses "the ethics and the aesthetics of participatory design." Our interpretation of PICTIVE ethics includes the empowerment of users to affect their work environments for the better. Our interpretation of PICTIVE aesthetics includes the practical demonstration that the enfranchisement of users can increase the enjoyment, interest, mutual education, and creativity in the design process.

ACKNOWLEDGMENTS

I am very grateful to Danny Wildman and Ellen White for their creative incorporation of what was then a largely unproven technique into our "Practicum for Methods in User Centered Design," and for their continued work with me to refine the technique. I thank Elizabeth Dykstra for many hours of discussion that have helped to determine the structure, use, and understanding of PICTIVE. I am also grateful to those designers and developers who have used the technique in their work: Kathy Cebulka, Peter Clitherow, Rob Farrell, Connie Kaye, Chris Koster, Larry Lefkowitz, Jay Messina, Todd Moyer, Gay Norwood, John Sauer Jr., and Danny Wildman. Thanks again to Danny and Ellen, and to Mary R. Smith, for serving as technical reviewers for this and related papers. I thank the following people for comments and discussions of the work: Sara Bly, Susanne Bodker, Elizabeth Brouckman, Rita Bush, Robert Carasik, Kathleen Carter, Tom Dayton, Joan Greenbaum, Saul Greenberg, Austin Henderson, Shu Hsi, Carl Lewis, Kim Halskov Madsen, Karen Murray, Rick Orriss, John Smith, Andrea Some, Carla Springer, Kari Thoresen, and Emilie Young. Randal Zoeller contributed his time by providing video operator services for some design sessions. I thank all of the participants for their enthusiastic participation in this ongoing experiment.

REFERENCES

Anderson, K. (1990). Micro-support staff plays key role in evaluation of new PC software. *PC-Week, 7*(23), 126.

Belenky, M. F., Clinchy, B. M., Goldberger, N. R., & Tarule, N. M. (1986). *Women's ways of knowing: The development of self, voice, and mind.* New York: Basic Books.

Bjerknes, G., Ehn, P., & Kyng, M. (Eds.). (1987). *Computers and democracy: A Scandinavian challenge.* Brookfield, VT: Gower.

Blomberg, J. L. (1990). Ethnographic field methods and their relation to design. In *PDC'90: Conference on Participatory Design* (p. ix). Seattle, WA. Palo Alto: Computer Professionals for Social Responsibility.

Blomberg, J. L. (1991). Improvising techniques to achieve participatory design goals: Everyone's challenge. In *Proceedings of CHI'91: Conference on Human Factors in Computing Systems* (pp. 389–390), New Orleans, LA. New York: ACM.

Bly, S. A. (1988). A use of drawing surfaces in different collaborative settings. In *CSCW'88: Proceedings of the Conference on Computer-Supported Cooperative Work* (pp. 250–256). Portland, OR. New York: Association for Computing Machinery.

Bly, S. A., & Minneman, S. L. (1990). Commune: A shared drawing surface. In *Proceedings of the*

Conference on Office Information Systems (pp. 184–192), Cambridge, MA. New York: Association for Computing Machinery.

Bodker, S., Ehn, P., Kammergaard, J., Kyng, M., & Sundblad, Y. (1987). A UTOPIAN experience: On design of powerful computer-based tools for skilled graphic workers. In G. Bjerknes, P. Ehn, & M. Kyng (Eds.), *Computers and democracy: A Scandinavian challenge* (pp. 251–278). Brookfield, VT: Gower.

Bodker, S., Ehn, P., Knudsen, J., Kyng, M., & Halskov Madsen, K. (1988). Computer support for cooperative design. In *CSCW'88: Proceedings of the Conference on Computer-Supported Cooperative Work* (pp. 377–393), Portland, OR. New York: Association for Computing Machinery.

Bodker, S., & Gronbaek, K. (1991). Cooperative prototyping: Users and designers in mutual activity. Submitted to *International Journal of Man-Machine Studies, 34*(3), 435–478.

Bodker, S., Gronbaek, K., & Halskov Madsen, K. (1990). Techniques for cooperative design. In *PDC'90: Conference on Participatory Design* (p. vi), Seattle, WA. Palo Alto: Computer Professionals for Social Responsibility.

Bonavoglia, A. (1991, April). Xerox participatory design project. Paper presented at CHI'90 workshop, *Participatory design: Practical stories and stories of practice*. New Orleans, LA.

Bruce, M., & Kirup, G. (1987). An analysis of women's roles under the impact of new technology in the home and office. In G. Bjerknes, P. Ehn, & M. Kyng (Eds.), *Computers and democracy: A Scandinavian challenge* (pp. 342–362). Brookfield, VT: Gower.

Bush, R. M. (1990). Putting the human in human-computer interaction. In *Proceedings of the National Communication Forum, 44.* Chicago: Professional Education International.

Carter, K. (1989). Two conceptions of designing. In S. Bodker (Ed.), *Proceedings of the 12th IRIS: Information Systems Research Seminar in Scandinavia,* (pp. 101–110). Denmark: Aarhus University.

Cebulka, K. D. (1990). WISE: An intelligent interface for user modification of applications. *Proceedings of 1990 IEEE International Conference on Systems, Man, and Cybernetics* (pp. 637–639). Los Angeles: IEEE Press.

Crane, D. (1990). Workshop on graphic recording in systems design. In *PDC'90: Conference on Participatory Design* (pp. 95–98), Seattle, WA. Palo Alto: Computer Professionals for Social Responsibility.

Crane, D. (1991, April). Changing the "demand" for participatory design. Paper presented at CHI'91 workshop, *Participatory Design: Practical stories and stories of practice,* New Orleans, LA.

Darwin, J., Fitter, M., Fryer, D., & Smith, L. (1987). Developing information technology in the community with unwaged groups. In G. Bjerknes, P. Ehn, & M. Kyng (Eds.), *Computers and democracy: A Scandinavian Challenge* (pp. 315–334). Brookfield, VT: Gower.

Dykstra, E. A. (1991a, April). Group design for and in shared workspaces. Paper presented at CHI'91 workshop, *Participatory Design: Practical stories and stories of practice,* New Orleans, LA.

Dykstra, E. A. (1991b). Practical requirements for participatory design. In *Proceedings of CHI'91: Conference on Human Factors in Computing Systems* (pp. 390–391), New Orleans, LA. New York: Association for Computing Machinery.

Dykstra, E. A., & Carasik, R. (1991). Structure and support in cooperative environments: The Amsterdam conversation environment. *International Journal of Man-Machine Studies, 34*(3), 419–424.

Ehn, P. (1988a). Playing the language games of design and use: On skill and participation. In *Proceedings of the Conference on Office Information Systems* (pp. 142–157), Palo Alto, CA. New York: Association for Computing Machinery.

Ehn, P. (1988b). *Work-oriented design of computer artifacts.* Stockholm: Arbetslivscentrum.

Ehn, P. (1990a). Statement for participatory design of computer systems. In *Proceedings of*

CHI'90: Conference on Human Factors in Computing Systems (p. 142). Seattle, WA. New York: ACM.

Ehn, P. (1990b, April). *Strategies and tools for participatory design.* Paper presented at PDC'90: Participatory Design Conference, Seattle, WA.

Ehn, P., & Kyng, M. (1987). The collective resource approach to systems design. In G. Bjerknes, P. Ehn, & M. Kyng (Eds.), *Computers and democracy: A Scandinavian challenge* (pp. 17–58). Brookfield, VT: Gower.

Floyd, C. (1981). A process-oriented approach to software development. In *Systems Architecture: Proceedings of the Sixty ACM European Regional Conference* (pp. 285–294). Guildford, Surrey: Westbury House.

Floyd, C. (1987). Outline of a paradigm change in software engineering. In G. Bjerknes, P. Ehn, & M. Kyng, (Eds.), *Computers and democracy* (pp. 191–210). Brookfield, VT: Gower.

Floyd, C., Mehl, W. M., Reisin, F. M., Schmidt, G., & Wolf, G. (1989). Out of Scandinavia: Alternative approaches to software design and system development. *Human Computer Interaction, 4*(4), 253–350.

Greenbaum, J., & Kyng, M. (1991). *Design at work.* Hillsdale, NJ: Lawrence Erlbaum Associates.

Grudin, J. (1988). Why CSCW applications fail: Problems in the design and evaluation of organizational interfaces. In *CSCW'88: Proceedings of the Conference on Computer-Supported Cooperative Work* (pp. 85–93). Portland, OR. New York: Association for Computing Machinery.

Grudin, J. (1990a). Obstacles to participatory design in large product development organizations. In *PDC'90: Conference on Participatory Design,* Seattle, WA. Palo Alto: Computer Professionals for Social Responsibility.

Grudin, J. (1990b). Statement for participatory design of computer systems. In *Proceedings of CHI'90: Conference on Human Factors in Computing Systems* (pp. 142–143). Seattle, WA. New York: Association for Computing Machinery.

Guba, E. G., & Lincoln, Y. S. (1989). *Fourth generation evaluation.* Newbury Park, CA: Sage.

Harrison, S., & Minneman, S., (1990). Design communications workshop. In *PDC'90: Conference on Participatory Design* (p. vii). Seattle, WA. Palo Alto: Computer Professionals for Social Responsibility.

Helander, M. (Ed.). (1988). *Handbook of human-computer interaction.* Amsterdam: North-Holland.

Holmqvist, B., & Madsen, K. H. (1989). Initiative in cooperative design. In S. Bodker (Ed.), *Proceedings of the 12th IRIS: Information Systems Research Seminar in Scandinavia* (pp. 271–292). Aarhus, Denmark: Aarhus University.

Kraut, R., Dumais, S., & Koch, S. (1989). Computerization, productivity, and the quality of work life. *Communications of the ACM, 32,* 220–238.

Kraut, R. E., Egido, C., & Galegher, J. (1990). Patterns of contact and communication in scientific research collaborations. In J. Galegher, R. E. Kraut, & C. Egido (Eds.), *Intellectual teamwork: Social and technological foundations of cooperative work* (pp. 149–172). Hillsdale, NJ: Lawrence Erlbaum Associates.

McKim, R. (1972). *Experiences in visual thinking.* Monterey, CA: Brooks/Cole.

Minneman, S. L., & Bly, S. A. (1990). Experiences in the development of a multi-user drawing tool. In *Proceedings of the Third Guelph Symposium on Computer-Mediated Communication.* Guelph, Ontario, Canada: CIPS.

Muller, M. J. (1991a). No mechanization without representation: Who participates in participatory design of large software products. In *Proceedings of CHI'91: Conference on Human Factors in Computing Systems* (pp. 389–392). New Orleans, LA. New York: Association for Computing Machinery.

Muller, M. J. (1991b). Panel: Participatory design in Britain and North America: Responses to the Scandinavian challenge. In *Proceedings of CHI'91: Conference on Human Factors in Computing Systems* (pp. 225–231). New Orleans, LA. New York: Association for Computing Machinery.

Muller, M. J. (1991c). PICTIVE: An exploration in participatory design. In *Proceedings of CHI'91: Conference on Human Factors in Computing Systems* (pp. 225–231). New Orleans, LA. New York: Association for Computing Machinery.

Muller, M. J. (1991d, April). Retrospective on a year of participatory design at Bellcore. Position paper presented at CHI'91 workshop, *Participatory design: Practical stories and stories of practice*. New Orleans, LA.

Muller, M. J. (1992). Retrospective on a year of participatory design using the PICTIVE technique. In *Proceedings of CHI'92: Conference on Human Factors in Computing Systems* (pp. 455–462). Monterey CA: Association for Computing Machinery.

Muller, M. J., & Cebulka, K. D. (1990). Software Professionals in the Year 2000: Technologies to support an enhanced social communications fabric. In *Proceedings of National Communications Forum*, 44. Chicago: Professional Education International.

Muller, M. J., Farrell, R., Cebulka, K. D., & Smith, J. G. (1992). Issues in the usability of time-varying multimedia. In M. Blattner & R. Danenberg (Eds.), *Interactive multimedia computing*. Hillsdale NJ: Lawrence Erlbaum Associates.

Muller, M. J., Kaye, C., & Sauer, J. E., Jr. (1991, September). *Case study of transformation of screen layout and dynamics through participatory design*. Poster presentation at 1991 Human Factors Society Meeting, San Francisco, CA.

Muller, M. J., Smith, J. G., Goldberg, H., & Shoher, J. Z. (1991). Privacy, anonymity, and interpersonal competition issues identified during participatory design of project management groupware. *SIGCHI Bulletin, 23*(1), 82–87.

Muller, M. J., Wildman, D. M., & White, E. A. (1992, May). *Games and other exercises for group design of user interfaces*. Tutorial presented at CHI'92, Monterey CA.

Namioka, A., & Schuler, D. (Eds.). (1990). *PDC'90: Conference on Participatory Design,* Seattle, WA. Palo Alto: Computer Professionals for Social Responsibility.

Naur, P. (1983). Program development studies based on diaries. In *Psychology of computer use.* London: Academic Press.

Nelms, H. (1957). *Thinking with a pencil.* Berkeley, CA: Ten Speed Press.

Open Software Foundation (1991). *OSF/Motif Style Guide, Revision 1.1.* Englewood Cliffs, NJ: Prentice-Hall.

Reason, P., & Rowan, J. (Eds.). (1981). *Human inquiry: A sourcebook of new paradigm research.* Chichester, UK: Wiley.

Schon, D. A. (1983). *The reflective practitioner: How professionals think in action.* New York: Basic Books.

Silverstein, G., Farrell, R., & Smith, J. G. (1990, March). Integrating multiple instructional paradigms in industrial training applications. *Proceedings of AAAI Workshop on Knowledge-Based Environments for Learning and Teaching.* Stanford, CA.

Smith, J. G., & Cebulka, K. D., (1990, April). *A framework for intelligent interfaces.* Poster presented at CHI'90. Seattle, WA.

Stefik, M., Foster, G., Bobrow, D. G., Kahn, K., Lanning, S., & Suchman, L. (1987). Beyond the chalkboard: Computer support for collaboration and problem solving in meetings. *Communications of the ACM, 30*(1), 32–47.

Suchman, L. (1988a). Designing with the user. *ACM Transactions on Office Information Systems, 6,* 173–183.

Suchman, L. (1988b). Designing with the user: Book review. *CPSR Newsletter, 6*(3), 1–6.

Suchman, L., & Wynn, E. (1984). Procedures and problems in the office. *Office: Technology and People, 2,* 133–154.

Tang, J., & Minneman, S. L. (1990). VideoDraw: A video interface for collaborative drawing. In *Proceedings of CHI'90: Conference on Human Factors in Computing Systems* (pp. 313–320). Seattle, WA. New York: Association for Computing Machinery.

Thoresen, K. (1990a). Prototyping organizational changes. In *PDC'90: Conference on participatory design* (pp. 22–33). Seattle, WA. Palo Alto: Computer Professionals for Social Responsibility.

Thoresen, K. (1990b). Statement for participatory design of computer systems (Panel). In *Proceedings of CHI'90: Conference on Human Factors in Computing Systems* (p. 144). Seattle, WA. Palo Alto: Association for Computing Machinery.

Verplank, W. (1990, April). *Graphical invention for user interfaces*. Tutorial presented at CHI'90, Seattle, WA.

White, E. A., Wildman, D. M., & Muller, M. J. (1991, September). *Practicum in methods for user centered design*. Tutorial presented at 1991 Human Factors Society Meeting, San Francisco, CA.

Whiteside, J., & Wixon, D. (1988). Contextualism as a world view for the reformation of meetings. In *CSCW'88: Proceedings of the Conference on Computer-Supported Cooperative Work* (pp. 369–376). Portland, OR. New York: Association for Computing Machinery.

Winner, L. (1980). Do artifacts have politics? *Daedelus, 109*, 121–136.

Winner, L. (1986). *The whale and the reactor: A search for limits in an age of high technology*. Chicago: University of Chicago Press.

APPENDIX 1: PICTIVE DESIGN OBJECTS

The following objects were obtained from conventional office supply vendors, and were used without modification:

Colored Felt-Tipped Pens. Colored pens serve as the participants' "keyboard." The pens should produce a thin line which is highly saturated with color, so that people can write without running out of space, and so that the video camera can easily record what they write. According to Crane (1990), there should be many earth-toned colors, not just black (too authoritative) and red (connotations of emergencies, warning, etc.) and yellow (connotations of high-lighting, and difficult to record on videotape). We have used pen sets consisting of 12 to 18 colors, and have observed participants making use of multiple colors in a systematic fashion. Colored pencils have not worked well for video recording: The lines they leave on the paper are not sufficiently saturated to show up on videotape.

Colored High-Lighters. The high-lighters are used to change certain data fields—i.e., as if the computer system were highlighting them, perhaps in a static (invariant) manner.

Colored Acetate. Acetate sheets—that is, suitable for photocopy transparency—have been very useful. For icon-based systems, the icons are provided on these sheets (see Appendix 2). Red or yellow sheets may also be cut up into field-sized or icon-sized pieces to simulate computer-generated dynamic highlighting.

Post-It Notes. Different sized Post-It notes have been used for pop-up menus (2 by 3 inches, or 3 by 3) or dialogue boxes (4 by 6 inches—grid-line-printed Post-It notes are useful in this size), as well as for participant-constructed icons (2 by 3, or 1.5 by 2 inches). It is often helpful to cut the Post-It notes into useful sizes—for example, oblong data fields or square icons—with a large paper cutter.

Scissors. Scissors help participants to tailor the materials to their needs.

Colored Constructed Paper. Colored construction paper could take the place of Post-It notes for large areas. We also used colored construction paper taped to large easel pads to serve as different sections of a printed report.

Removable Tape. Removable tape helps to position and reposition objects that are larger than Post-It notes—for example, construction paper sections.

Colored Erasers. The colored erasers were not actually used. However, their bright colors and boxy shapes added to the overall perceptual quality of the experience (see the earlier section on visual communication).

These objects were arranged on a desk blotter pad, permitting users to match objects via tape or glue, to draw connecting lines or to draw new objects or representations directly on the pad.

APPENDIX 2: CONSTRUCTING THE DESIGN OBJECTS

As in the work of Bodker et al. (1987), we prepared paper-and-plastic versions of certain elements of the target computing system.

Icons. Plastic icons for graphical user interfaces were produced in several steps:

1. A graphical drawing program was used to develop images for the icons. These were usually much larger than the eventual icons would be on the target system—but their size made them more "play-like" (see the earlier section on Design Games, and Bodker et al., 1991) and encouraged participants to re-design them.
2. The images were printed onto paper. Each sheet of paper could hold the complete set of icons, to facilitate comparison and discussion.
3. The paper was then copied xerographically onto acetate sheets. The icons were cut out of the sheets, and included among the design objects.

Pop-Up Events. Pop-up events were produced in several steps:

1. A graphical drawing program was used to develop images of the pop-up events.
2. The images were printed to paper.

3. The paper was photocopied and cut, to produce a sufficient number of identical events so as to encourage their use as "throw-away" item, suitable for modification by the participants.

Menu Bars and Other Fixed Elements. Menu bars and window frames and the like were constructed similarly to the method for icons.

Paper Reports. When PICTIVE was used to design paper reports, we used a large easel pad with a one-inch grid printed onto its pages. Different sections of the report were drawn in outline onto the pad. Colored sheets of construction paper were cut out to match the size and shape of the outlined sections, and were provided in sufficient quantity to encourage their use as "throw-away" items, suitable for modification by the participants.

11

Reciprocal Evolution as a Strategy for Integrating Basic Research, Design, and Studies of Work Practice

Christina Allen
Northwestern University

Picasso is a collaborative project for research on the use of multimedia communications channels for supporting co-work; in particular, for research on how they change, enhance, and detract from communications and work practices.

Philips N.V. Corporation initiated the Picasso project because they found that the learnability and usability of their technologies increasingly was an issue in the success of their products on the market. Philips has a massive market presence in homes, on the road, and at work, where people use technologies such as telephones, facsimile machines, answering machines, intercoms, computers, radios, televisions, VCRs, and CD players in their daily activities. As technological developments afford the possibility of more and more features, these devices become increasingly difficult to learn and use. Moreover, with technologies whose use affects such existing social conventions as those of communication and work relationships, people more and more frequently find themselves engaged in activities for which there are no well-established conventions or genres of use. Until conventions of use become stabilized through their social construction, technologies are particularly hard to understand. Philips believed that learnability and usability issues had the potential to drive the future competitiveness of their products, and were frustrated with available approaches to the study of technology learning and use. Thus, the Picasso project involved an exploration of current concepts of information, communication, and learning, as well as methods for integrating this understanding into the design, development, and support of products. This demanded a principled understanding both of the nature of human communication and learning in real settings, as well as of the relationship between the design and adaptation of multimedia communication technologies. To respond to these needs, a method called *reciprocal evolution* was developed for use within the project, and subsequently by Philips. In this paper, I characterize reciprocal evolution, and discuss both its relationship to other methods and the implications of the method for corporate design and research efforts.[1]

RECIPROCAL EVOLUTION

The key idea underlying reciprocal evolution is that *use is design*. Technologies are commonly used in ways different than those envisioned by their inventors and designers. As people integrate a device into their work environments, their emergent patterns of use determine what the product *is* for them. For example, the now-dominant use of computers as word processors was unimaginable to early computer manufacturers, who conceived of computers as mathematical machines. How might we leverage the design revealed by human use of technologies?

Even when individuals work alone, the concepts they use are the result of social, collective, and constructive processes that are shared and distributed throughout communities. Technologies function like concepts, as interpreted cultural artifacts. Recognizing this fact, reciprocal evolution is based on the understanding that the quality of a technology is not determined solely by technical factors or by considerations of cognitive analyses of tasks or users, but substantively by the social organization of work and community around learning and using the tool in real-world work activities.

Technologies are thus thought of as providing *data collection points* for a deeper understanding of how a particular tool and tool type is used—not only in the communities of use, but in the larger evolving cultural definition of the social and work values of tools.

The leverage for improving technologies lies in deepening our understanding of the human design activities found in work practices with technologies. How do people adapt their existing practices to integrate new technologies, and how do they adaptively use and change technologies to support their existing and desired practices? To exploit the insights available in recognizing that *use is design* we found we must look at what users do in their work settings, not only ask them about their practice, or study their reactions to technologies in laboratory situations.[2]

In sum, we were concerned to create replicable methods for building and

---

[1]For additional information about the research and design findings of the Picasso project, see Allen, Linde, de Vet, Pea, & de Vogel, 1991; Linde, in press.

[2]Kling (1987) highlights the importance of social and organizational factors for understanding technology implementation successes and failures, but relies primarily on social science techniques such as interviews.

studying systems that evolve, not only in response to new technological capabilities, but to design opportunities emerging from continual study of how current products are changing people's organization of work and interaction, and their expectations and desires for technology functionality.

Component Activities of Reciprocal Evolution

Our concept of reciprocal evolution has roots particularly in Gregory Bateson's work on communication, and his development of concepts from cybernetic theory for human interaction and cultural learning (e.g., Bateson, 1972). This approach integrally depends on creating a cybernetic research and development system in which the outputs from various activities—technology design, work practice and learning, and basic research—can feedback and feedforward into reconstituting the futures of each of those activities. Results of these inquiries may then be exploited in new prototyping activities, research programs, or changing workplace studies.

In our studies of emergent work practices with new technologies, we often found recurrent problems, innovative solutions to perceived needs, and dilemmas in computer-enhanced collaborative work rich with design implications, and whose structure and processes illuminate basic research on the organization of talk and activity in the social sciences. We were concerned to establish a balance of disciplinary contributors to our work, as well as a set of activities that would contribute to the continual development in these three domains of reciprocal evolution: study of work practices with technology, the design of technologies, and furthering basic research. These three activities are described in greater depth next.

Study of Work Practices.   This category of activity focused on a particular set of research sites, with whom we have long-term relationships. We were interested in understanding how people reorganized their work practices and relationships when changes occur in the work environment. We studied the role that the technologies assumed in their day-to-day work and interactions, how the technologies were used in different ways than the designers intended, and what new organizations of work and communication were made possible or obvious by the tools.

It is important to understand that individuals and work groups continually reshape their activities out of the resources that they have available (Allen, 1988; Brown, Collins, & Duguid, 1989). Given that in most modern work settings, conditions are prone to rapid change, we were concerned with the methods and premises by which things changed, and the learning and uses of tools that evolved over time in these settings. Thus, our methods had to expose these differences in perception of appropriate learning, use, and work practices by individuals, local work groups, and institutions.

Design Implications.   We designed specific technologies for use in sites, and also determined generally important implications of the research for related standards, platforms, and operating systems used by Philips. The design implications that emerge from the basic research and the study of work practices describe general expectations for how a technology designed in a particular manner will succeed or fail, and why. The technology must also be thought of as a probe with which to investigate and stir up work practices, as well as to suggest new research directions.

Beyond technology design and standards implications, we determined aspects of the context for the use and learning of the technology, methods of introduction, and support and training for ease of use and learnability.

Basic Research.   We drew from basic research in fields contributing to an understanding of everyday and work-practice cognition and learning. These fields included cross-cultural cognitive psychology, cognitive anthropology, learning studies examining discourse practices (e.g., McDermott, 1982), anthropological linguistics and conversational analysis (e.g., Goodwin & Heritage (1991)), technology design (e.g., Carroll, 1990; Ehn, 1988), and ethnography. Reciprocal evolution is both guided by research and contributes to it, as we elaborate basic research findings by examining their relevance to our observations of technology use and learning in work settings.

Much of the basic research that has influenced our work has investigated how people communicate and work together (e.g., Sacks, Schegloff, & Jefferson, 1974). However, in the main it has not looked at how people communicate and work with one another in a technology-rich environment. In this area, we have extended the research with a focus on the effect of technology on cowork as well as the nature of interaction around tasks (e.g., Linde, in press).

Our Sites and How we Study Them

Our focus was on small business fieldsites with minimal to medium-complexity computer tool use, and not, for example, on technical/engineering work settings such as university departments, manufacturers, or large businesses. A major reason for this emphasis was the shift toward desktop markets for computing and communications applications, moving beyond a technically literate user population to the business community at large. For this reason, the prototype program used in the final phases of our Picasso research work (Picasso 2.0) integrated the functionalities of telephoning, faxing, scanning, electronic file transfer, videophone, and screen sharing. One field-work site was GG, a graphic design company with two principals who live and work 50 miles apart. They have been using various Macintosh applications in their work for several years.

HELP, our second site, is a nonprofit private welfare agency that builds and manages housing for the physically and developmentally disabled and supervises their case-workers. Their work is spread across five building sites, and includes coordinating reports on their clients, mortgages of their properties, private and public financial accounting, and coordinating research for these activities and reporting with their coworkers, government agencies, and banks.

Ecologically Valid Data.   The studies took advantage as far as possible of naturally occurring situations in which the technology was used. Our data on work practices included ethnographic video records of work with technologies during ordinary work tasks in real worksites, in addition to participant observation and ethnographic interviews. These interviews asked open-ended questions of participants about the nature of their work, and its objects and processes, without asking for accounts (rationalizations) about *why* they do what they do.

Study of Current Practice Before Tools are Added.   We studied the changes in work practice that emerged as new communication and computational technologies were integrated into the work sites of practitioners. Through long-term relationships with the sites, we were able to examine the practices in place before the tool comes into play, the character of tool learning and work practice adaptation for individuals and groups, and changes in work practices and organization of work over time. This strategy allowed one to understand the evolution of work practices before and after the technology was introduced. It provided this information in real time so that it could affect the design of future generations of technology. We used such a comparative method because current interactions provide for users the "standard" case, in contrast to which they evaluate a new technology. In addition, we needed to understand for tool design the features of the current form of human interaction for a given cowork task.

Data Collection in-situ with Multiple Methods.   We used a broad set of methods for collecting data, because various types of data were needed for understanding the coevolving nature of work practice and multimedia communication tool design. But it was also possible to gather by means of structured interviewing instruments more opportunistic information on how such a piece of technology became "encultured" as an unremarkable and routinely used object in the social and cultural texture of cowork. We analyzed these materials using methods of conversation and interaction analysis (e.g., see Goodwin & Heritage, 1991). We investigated the local reasoning spontaneously used by participants to ground their actual work practices.

Qualitative Inductive Research.   The appropriate research design for our studies was a qualitative inductive rather than a quantitative hypothesis-testing framework. Hypothesis testing is an appropriate research strategy when much is known about the phenomenon of interest, so that the crucial variables can be identified. For the Picasso studies, what was at stake was precisely the identification of the relevant variables, that is, the resources and constraints whose presence shapes people's interaction with technology. Because there were not any ecologically oriented theories about such activity, a rigid hypothesis-testing approach would have been an inappropriate choice.

Design and Research in Combination

Just as the social scientist brings particular expertise to the establishment of research sites and the study of interaction over time, so do the designers bring expertise about technological options, innovations, and how to embody research findings into workable designs. The perspectives of the designers and the scientists are complementary, yet different. It is rarely possible for either one to interpret the data collected from the sites for the other. Each sees in the studies different information valuable to their area of concern, and each comes to inculcate the other into seeing from a new perspective, and with different goals. For example, the goal of the scientist is to discover and articulate, the goal of the designer is to understand and transform that understanding into useful forms. Thus, it is crucial that these disciplines work closely together in all aspects of the process, yet often with one taking the lead and doing the bulk of the work in their area of expertise. It should never be the case that one "pre-digests" data or designs for the other.

In our case, within the constraints of the 3-year funding period available, we were able to proceed through two iterations of design and implementation. We began with studies of work practice in our work sites before our technologies were introduced, designed, and developed a first version of the prototype tool in situ for one work organization (GG), and then on the basis of our work in the three areas of reciprocal evolution, designed and created a second version of the tool for use in the GG and HELP sites. Our final report to Philips articulated findings of coworker learning and use of Picasso 1.0 and 2.0, and speculated on promising design directions for Picasso 3.0.

RECIPROCAL EVOLUTION IN RELATION TO OTHER METHODS

This section discusses the relationship between reciprocal evolution and other methods, such as an engineering perspective, a cognitive science perspective, and related work using participatory design methods, in order to understand the relevance of these various approaches to different aspects of designing and different contexts for design.

Basic science and design work are often the polar pulls of research on human-computer interaction, and a broad range of relations between these poles is represented in the systems design and human-computer interaction communities.

More rarely do observations of work practices figure in the strategy for relating science to design, as we emphasize in reciprocal evolution. Yet there is very much more to the evolving functions of technology in work practices than the cognitive interface between the computer and the "user" at one extreme, or the system design perspective driven by institutionally defined work task requirements, at another extreme. In each of these extremes, the machine plays too large a role in the analysis, when in fact it is but one artifact shaping the work life of workers and communities-of-practice.

Reciprocal evolution seeks not universal principles as in science, nor single particulars as in participatory design, but *design dimensions* from recurrent problems emerging out of the interdisciplinary microanalyses of videotaped records of situated work practices with tools (Allen, Linde, de Vet, Pea & de Vogel, 1991). We argue that these dimensions provide fertile invention "materials" for new tools as well as salient themes for looking at work practices with tools other than the specific ones that gave rise to the dimensions in our analyses.

Reciprocal evolution is a strategy that has general applicability in sites where participatory design is not possible or where broader design implications are sought from user-centered system design. Worksite observations across locations for how these technologies are used may provide both specific and general information about the work practices and the specific technology.

Engineering Perspective

The reciprocal evolution perspective is particularly distinctive from the usual engineering orientation central in many industrial development laboratories. In such an approach, engineers look at the new functionalities made possible by the technical capabilities of new computers and communication devices, and, with little consideration of the contexts of use and work practice, create designs and products that pile feature upon feature. Telephone answering machines and VCRs are among the most dramatic examples of the engineering orientation, and have unfortunately often resulted in tools whose functionality is available only to a small subset of people with the patience to learn the complexities of such systems.

Cognitive Science Perspective

We find some overlap between reciprocal evolution and Carroll's (1990) approach to minimalist instruction for practical computer skill, especially in his emphasis that designers need to work much more with what learners do spontaneously to find meaning in their learning activities with new systems. However, our approach to design differs from Carroll's in two fundamental respects. First, whereas Carroll focuses on the first learning of new tools, we are concerned with continuing learning of users over the life cycles of their work tasks, and continuing redesign. Second, whereas Carroll's approach is fundamentally a laboratory-experimental paradigm, we find it essential to observe the rich resourcefulness of learner use and social interactions around use of technologies in their situated work practices within institutions.

It is a working assumption of ethnomethodology (an approach to microsociology developed by Garfinkel, 1967 [also see Heritage, 1984] that informs our work) that the participant is not a privileged informant to his or her own activities. For this reason, we do not ask for think-aloud protocols, although we take the utterances and explanations by the participants that arise during their work activities as evidence, along with other observables, of the sensible nature of the activities under observation, from the participants' perspectives.

We also find the theory-based design guidelines for human-computer interfaces offered by Newell and Card (1985) problematic. Although their emphases on the hierarchical structure of goal-oriented behavior in human interaction define a paradigm aimed at leading to efficiently fast and low-error human performances, these information processing system considerations are too impoverished to suggest promising directions for design and invention of tools in the first place. Their work also ignores the creative contributions of users in making sense of their tasks and improvising novel methods and unique task decompositions to do their work.

Participatory Design and Related Social Science Research

The participatory design method under development by several research groups is designed to address many of the inadequacies of engineering and cognitive science perspectives. It has the strengths of examining technology in real work practices, and in such pioneering work as that of Ehn and Bødker (e.g., Ehn, 1988), of jointly envisioning future workplace scenarios by collaborative teams of workers, systems designers, management, and union. It also has the important feature of providing concrete examples for design successes arising from the application of participatory design processes (e.g., Allen, in press).

However, participatory design is and has been designed as a method primarily for fine-tuning specific technological tools for particular work environments, and with significant union involvement (e.g., Bjerknes, Ehn & Kyng, 1987; Ehn, 1988; Suchman, 1988). We ask where the evidence is for how this collaborative process works from the point-of-view of the different participants in the "collaboration"? In other words, how is the democracy now evident in the behaviors, beliefs, etc. of the participants after the designer leaves? Who manages, updates, rewrites the technology without the impetus of democracy-minded computer scientists? There must be an analysis of the distinctive pressures on each of the parties to shaping the collective work they are doing in participatory design. How is the existence and the collaboration between the designers and the shop floor employees any different than any other institutionally instigated event that takes place for them?

The social work of collaboration on the part of participatory designers with workers paves the social way for the introduction of the new technology, but in our opinion would be unlikely to generalize to other sites. It is thus removed from the scientific objective of seeking out general patterns. It is directed to solving the design problems for the specific case, and leads to the relative abandonment of a search for generalities, which have application across sites and across specific product designs.

Lave and Wenger's (in press) work focuses on trades such as butchering and tailoring that are thousands of years old and not significantly changed over that period. However in America, as in other countries moving away from agriculture and industry as the primary trade base, trades come and go at a rapid pace. And people are coming and going as well—general education, a transient population, and a diminishing of long term corporate employment has resulted in new forms of work and learning that are far from the traditional apprenticeship opportunities imagined by Lave and Wenger or extant in the newspaper rooms and hospitals of the participatory design literature.

Within Philips, and for many other tool designers and builders, technologies are designed generally, rather than specifically. Thus it is hard to say that participatory design will cover the majority of the tools designed, or even a large minority, since the majority of tools are rarely designed for specific workers, institutions or work sites. It is for these reasons that we have focused on methods for drawing the general out of the specific, and have coupled that with a continuous look at what is happening to our tools as they are appropriated and used in real work settings.

INDUSTRIAL IMPLICATIONS OF RECIPROCAL EVOLUTION

The ultimate motivation for a commercial company to start and support projects such as Picasso is to develop, introduce, and support commercially successful products. In the development of this method, we were concerned with both design and research, but also with the industrial applications of such findings. The practice of reciprocal evolution has implications for how corporations such as Philips think about their markets, and how they organize their product design and development cycles. In this section, we review its consequences for the strategies and activities of the corporation.

Studying Real Settings, Users, and Communities

One central conclusion from the Picasso Project has been the growing recognition of the necessity of obtaining "ecologically valid data" in the research one carries out. Although there is indeed value in some aspects of experimental science involving new technologies with humans, such as in ergonomics, vision, and audition, the lessons learned from such investigations are but a piece of the much larger puzzle of how human beings relate to technology and give it meaning in their daily activities. And even such experimental studies as these may be invalid if they are based on assumptions about what the workplace "must" be like, rather than on actual observation of the workplace.

Learning is an obvious case in point. Most people learn a new technology because they want to use it to do their own work, rather than because they want to master the entire technology. They sense a gap between their desires and the current structure of their world, and perhaps, they reason, the tools can be used to narrow that gap. To find out, they may struggle to connect the tool and their work tasks, and their inventive tool uses often turn out to be very surprising to the designers. It is important to note that the way naive users use a technology is never just an incomplete version of the way expert users use it. There is often an integrity and unexpected success in the use patterns that a "naive" user has created for tools in his or her community.

Furthermore, the environment in which learning often takes place is a complex mess. Workers rarely learn in soundproof white rooms. They learn amidst other tasks, ongoing responsibilities, multiple conversations, and the participation of coworkers who move in and out of the learning process. We have also recognized that every worker at a site does not have to be an expert user of the resident technology. Rather, people know who else in the office has expertise with a given technology, and who of those people can be asked for help. We sought to determine how this emergent and distributed expertise was integrated or excluded from the institutional view of "expertise," and "responsibility" for using, learning and teaching others about the technology.

Orr's (1990) recent ethnography of a modern job involving "talking about machines" contributes evidence of the mis-match of relations between emergent organizations and institutional expectations. His research demonstrates how technical representatives of a major copier company have to work *around* the corporate policies and training in order to get the expertise that they need to do their jobs. The corporation's beliefs about the expertise needed to fix a copier, and the abilities of the technical representatives to comprehend certain kinds of information, leads the corporation to design training that is inadequate for the tasks facing these representatives in customer sites. By defining alternative ways of gaining expertise (telling "war stories") in the absence of organizational support or understanding, the technical representative are able to succeed in their work, but the expectations of each group continue to grow apart, leading to greater potential for conflicts and tensions.

Finally, we are interested in how peoples' roles and conveyed attitudes contribute to the pattern of learning and use that takes place in organizations. These issues are important because they merge with issues of staff development and worker-union concerns, providing important information that should influence organizational design, corporate planning, and corporate strategy.

Interdisciplinary and Interdivisional Efforts

We believe that it was critical to have the different members of our team look at data together. It is principally through these communications that the team members conveyed the diverse design, social, and cognitive perspectives that underlie the project work.

In large projects, with people joining and leaving over the years of development, the expected characteristics of the future users of a technology are either never explicitly stated or are as diverse in profile as are the team members. The type of data collection suggested by reciprocal evolution offers opportunities for consolidation and consensus building during the design process. We have come to recognize that video is a rich medium for enriching design because video records can focus discussion for researchers and designers, and can afford a basis for conversation far more meaningful than abstract descriptions in written language (cf. Suchman & Trigg, 1991). Each participant still has his or her own expertise, but converses with everyone else during work practice video analysis about what they perceive to be going on.

We believe that it is important to establish tight links between product development and research groups in order to get a better understanding of technology in use, and the work practices of market cultures in general. Every group involved in product development, from marketing, to design, to engineers, to salespeople must be involved in the process of determining what it is that people actually do with products. It is not enough for research to study these users and inform the designers because it is from the multiple perspectives on these data that useful insights and interpretations about the meaning of users' work practices can emerge.

Early Investigation and Long-Term Studies

An important extension of this project would be to apply this way of working to even the earliest step of product development, the *concept development phase*. If future users of a technology are involved in the conceptual part of the development of new devices or systems, it remains on the abstract level as described earlier in "testing." We see work now fairly limited in its use of questionnaires, set interview schedules, and checklists. Rightly used, this phase could form a departure point for development; workers and designers could engage in significant joint envisioning of work tools in the contexts of work practice.

These investigations need to occur over a protracted period, and they must begin before the introduction of new technologies to a work site, so that the transformations of work practice that arise can have a baseline comparison. This implementation would involve increased involvement with the relevant groups in the corporation as they work to develop and design advanced technologies. On a project basis one could incorporate the reciprocal evolution method, augmented with other design methods that guarantee user involvement and feedback. While this process is a significant departure from fairly compartmentalized groups with delineated responsibilities in part of the linear model of product development and design we see today, we see the likely payoffs for a more dynamic and innovation-sensitive organization to be worth the efforts of invention.

Changing the Design of the Product and the Organization

One purpose of these studies of work practices is to produce more learnable and usable technologies for real work settings. A good part of our emphasis has been to seek a way to build into the research and development process new opportunities for seeing what customers make of the design innovations arising in the corporation. This involves seeing how users go on designing in their very use of a product, how they construct new conventions for using technology in their work, and how they evolve new conceptions of work practice that can give rise to further rounds of technological innovation. What we imagine is design iterations that seek to learn opportunistically from the marketplace in a deep sense, where the users are viewed as coinnovators in their adaptation of tool to work, and work to tool, and not just for the design process, but for marketing, strategy, customer support and service.

Support services need to begin a conversation with users by seeking to understand what the users have *made* of the technology in their work, how they have been using it, and what they wish to do with it. The corporation could be exploiting these conversations as a form of market research—users actually are providing an important window through such support calls into how they see the product and its usability for their intended work. This further implies that a link be forged between these support staff and the designers of technologies.

As part of this commitment to seeking an understanding of evolving design through use, there needs to be a corporate commitment to modular approaches to product development, so that insights from real users can be incorporated quickly into the next versions of products. This means that there must be a use of rapid prototyping and high level prototyping languages so that ideas can be tested with real users in real work settings before new concepts become products. At a larger level, the corporation should be studying the use of already introduced products as a basis for understanding general ways in which the product better could be designed or supported.

Changing the Basis of Corporate Long-Term Planning

Too commonly, research is viewed as a cost and not an investment. We argue that to be competitive in the long term, a business should think of a single product not as a finished entity, but as a punctuation point in an ongoing cycle of design, based on attention to emerging needs and desires which foreshadow future directions of existing markets and the potential for new markets. While market research and other forms of "probing" the user have rested on the same intuition—seek the future by looking at emerging trends and beliefs today—the nature of market research methods relies too much on reflection and abstraction outside the contexts of work in which people make sense of their technologies, missing much that is going on. The method of reciprocal evolution lives in the place of work, just as the tools that are evolved through design and use must.

A corporation should be committed both to the technology and to the market, with market-driven plans for technical development in the long term, and with technological innovations widely tested in the market before long-term directions for a product line are established. This requires participation and support for such pilot projects from the very top, and quick successes to show immediately. There is not so much a need to convince the research community, but rather the decision makers in the product development and marketing groups in corporations (cf. S. Zuboff's commentary on John S. Brown's article "Reinventing the Corporation," Harvard Business Review, 1991: "there is no use in convincing researchers, management is the decisive factor if you want to gain influence"). Because working with this approach means a radical shift in the whole design process level, and a different way of looking at customers and markets, the introduction of such methods requires quick successes to gain momentum.

SUMMARY

New technologies are transforming the work of individuals and organizations throughout the world. But too commonly it is principally a technology-centered approach that drives the design of tools we see in the hands of workers. In the development of a dynamic interdisciplinary method for continually integrating basic research, design, and studies of work practice, we have begun to forge an alternative perspective that respects the fundamental design, which users contribute to these tools as they learn and use them in their activities. Throughout a pilot project with a major corporation, real research sites, and a research institute, we have investigated a new form of product development based on this integration. Our experience indicates that reciprocal evolution holds significant promise as a continual and integrated part of corporate strategy, research and design.

ACKNOWLEDGMENTS

Please address correspondence to the author at The Institute for Research on Learning, 2550 Hanover Street, Palo Alto, CA 94304 USA. The Picasso Project was funded through general sponsorship of IRL by Philips N.V. Corporation. Collaborators include Charlotte Linde, Roy Pea, Rob de Vogel, and John de Vet. "We" in the report refers to the collaborative voice of this research team.

REFERENCES

Allen, C. (1988). *Situated designing.* Master's Thesis, Carnegie Mellon University, Design and Computing.

Allen, C. (in press). Multimedia learning environments designed with organizing principles from non-school settings. In E. de Corte, M. Linn, H. Mandl, & L. Verschaffel (Eds.), *Computer-based learning environments and problem-solving* (NATO Series, Subseries F: Computer and System Sciences). New York: Springer-Verlag.

Allen, C., Linde, L., de Vet, J., Pea, R. D., & de Vogel, R. (1991). *The Picasso Project* (IRL Tech. Rep.). Institute for Research on Learning, Palo Alto.

Bateson, G. (1972). *Steps to an ecology of mind.* Ballantine Press.

Bjerknes, G., Ehn, P., & Kyng, M. (1987). (Eds.). *Computers and democracy: A Scandinavian challenge.* Brookfield, VT: Avebury.

Brown, J. S., Collins, A., & Duguid, P. (1989). Situated cognition and the culture of learning. *Educational Researcher, 18*(1), 32–42.

Carroll, J. M. (1990). *The Nurnberg Funnel: Designing minimalist instruction for practical computer skill.* Cambridge, MA: MIT Press.

Ehn, P. (1988). *Work-oriented design of computer artefacts.* Stockholm, Sweden: Almquist and Wiksell International Publishers.

Garfinkel, H. (1967). *Studies in ethnomethodology.* New York: Prentice Hall.

Goodwin, C., & Heritage, J. (1991). Conversational analysis. *Annual Review of Anthropology, 19,* 283–307.

Heritage, J. (1984). *Garfinkel and ethnomethodology.* Cambridge, England: Polity Press.

Kling, R. (1987). Computerization as an ongoing social and political process. In G. Bjerknes, P. Ehn, & M. Kyng (Eds.), *Computers and democracy: A Scandinavian challenge* (pp. 111–129). Brookfield, VT: Avebury.

Lave, J., & Wenger, E. (1991). *Situated learning: Legitimate peripheral participation.* New York: Cambridge University Press.

Linde, C. (1991). What's next? The social and technological management of meetings. *Pragmatics, 1,* 297–318.

McDermott, R. P. (1982). Institutionalized psychology and the ethnography of schooling. In P. Gilmore & A. Glatthorn (Eds.), *Children in and out of school.* Arlington, VA: Center for Applied Linguistics.

Newell, A., & Card, S. K. (1985). The prospects for psychological science in human-computer interaction. *Human-Computer Interaction, 1,* 209–242.

Orr, J. E. (1990). *Talking about machines: An ethnography of a modern job.* Doctoral dissertation. Cornell University.

Sacks, H., Schegloff, E. A., & Jefferson, G. (1974). A simplest systematics for the organisation of turn-taking for conversation. *Language, 50,* 696–735.

Suchman, L. (1988). Designing with the user: Review of *Computers and democracy: A Scandinavian challenge. ACM Transactions on Office Information Systems, 6*(2), 173–183.

Suchman, L. A., & Trigg, R. H. (1991). Understanding practice: video as a medium for relection and design. In J. M. Greenbaum & M. Kyng (Eds.), *Design at work: Cooperative design of computer systems* (pp. 65–89). Hillsdale, NJ: Lawrence Erlbaum Associates.

IV CASE STUDIES

12

The Participation of Users in Systems Design: An Account of the Origin, Evolution, and Use of the ETHICS Method

Enid Mumford
Manchester University, England

The word ETHICS is an acronym for Effective Technical and Human Implementation of Computer-based Systems. It is also an expression of a personal philosophy. I believe that the future users of new technical systems should be able to participate in the design process and help create systems that are humanistic and friendly as well as efficient and effective. My interest in changing systems design practice was stimulated by observing the bad human effects of many early computer systems. Work was frequently routinized and controls tightened as a result of the new technology. Systems analysts, as designers were called then, appeared to have little understanding of the human consequences of their work. The difficulties of technical design appeared to displace any concern for human feelings.

When computers first appeared in offices in the late '50s and '60s their costs and limitations meant that they were often introduced in an authoritarian manner. "This is what we can provide and you must have it" was a common technical attitude. Then, as user resistance was encountered, strategies changed to a soft sell approach: "This is what we can offer and it is just what you want." Overselling of poor systems led to user skepticism and gradually analysts began to realize that they needed to talk to users before producing a product: "We think we know what you want but we'd like to discuss this with you." This led to the practice of interviews with single users.

Today, a quasiparticipative strategy is increasingly used: Technologists are saying: "We think we know what you want, but we'd like you to help us design the system to make sure we get it right." This usually requires users to spend time defining their information needs. It is an approach that can pay off, as a clear and correct specification of information needs is a critical factor in the successful design of information systems. ETHICS takes participation a stage further. Users are asked as a group to identify the problem that they would like a computerized system to tackle; to specify their information needs; to set objectives for the new system; and to redesign departments and jobs so that the total system can work effectively (Mumford, 1983a). The design task is shared with the technical designers of the system who will concentrate on the provision of appropriate hardware and software.

WHAT IS PARTICIPATION?

Because ETHICS is a highly participative approach it is useful to discuss briefly the meaning of participation. It is not a new concept although it has been given other names such as democracy, involvement, sharing, cooperation. The Greeks used it to describe a certain kind of decision making. A decision was democratically taken if the answer to the question "who takes it" was "more or less everybody." This kind of democratic decision taking contrasts with meritocratic in which decisions are taken by an elite, and with autocratic in which they are taken by one man (Lucas, 1976). Democratic or participative decision taking has two other important components. Decisions are reached by discussion, criticism, and compromise. They are also seen has having as their principal objective the interests of all, rather than those of a single group, faction, or party.

A definition of participation that fits the philosophy of ETHICS is the following:

> Participation is a process in which two or more parties influence each other in making plans, policies or decisions. It is restricted to decisions that have future effects on all those making the decisions or on those represented by them. (French, Israel, & Aas, 1960)

The arguments for participation when new technology is being introduced are varied and complex. Companies may use a participative approach because they believe that participation is morally right—people should be able to influence their own destinies. Or they may see participation as helping to produce an interested and committed workforce and therefore assisting in the avoidance of industrial relations problems. They may agree with John Stuart Mill that participation is a valuable educational experience and provides understanding and knowledge that can assist an organization to more effectively realize its objectives or, alternatively, that such knowledge will enable any bargaining that takes place to do so with all interest groups operating from an informed position. These are all positive reasons for participation. It is clear, however, that some managements favor participation for negative reasons. They see it as a means of persuading their employees to accept changes that otherwise might be rejected. Participation then becomes a manipulative tool and is carried out without any serious attempt to make it truly democratic.

Trade union arguments for participation can be equally varied. Unions may encourage participation because they see it as a lever for increasing shop floor control over the work situation. They may also support it because they believe it will contribute to industrial democracy and the more equitable sharing of power. But some unions reject participation, stating that it is their responsibility and right to represent the interests of their members and that they can do this effectively without involving these members.

WHAT IS ETHICS?

ETHICS is intended to provide users who are not technologists with the means to control or influence systems analysis and design. It does this by involving them in the design processes and providing tools and techniques that assist an analysis of their needs and problems. This enables them to design the system from an organizational perspective (Maclaren et al., 1991). At present all ETHICS tools are paper based.

ETHICS has three principal objectives. These are:

1. to enable the future users of a new system to play a major role in its design and to assume responsibility for designing the work structure that surrounds the technology. This involves a learning process and a set of simple diagnostic and sociotechnical design tools.

2. to ensure that new systems are acceptable to users because they both increase user efficiency and job satisfaction.

3. to assist users to become increasingly competent in the management of their own organizational change so that this becomes a shared activity with the technical specialists and reduces the demands on scarce technical resources.

The methodology is not necessarily aimed at producing a computer-based solution as the emphasis is on obtaining the right balance between the social and technical aspects of the complete system. ETHICS incorporates the joint philosophies of participation and sociotechnical design. The following statement expresses my own definition of the design task (Mumford, 1983b):

> Good organizational design requires a decision STRUCTURE that incorporates all interested groups affected by the new system; a PROCESS which enables the design task to be smoothly carried forward from identification of need to change to successful operation of the new system, and a CONTENT that enables business efficiency and employee job satisfaction improvements to be considered in parallel and given equal weight.

ETHICS incorporates the following diagnostic and design tools:

1. A framework to assist the identification of mission, key tasks, and factors critical to effective operation.
2. A questionnaire to measure job satisfaction.
3. A variance analysis tool to assist the identification of problems and problem areas.
4. A framework to identify what is likely to change in the internal and external environments.
5. A set of guidelines for individual and group work design.

A user group designing an information system with ETHICS would proceed through the following steps:

A. Mission and Key Tasks Description.—of the department or function where the new system will be located.

At the start of the design process the design group will spend some time working out clear answers to the following questions:

1. Why does the department exist, what it is trying to achieve?
2. What important tasks must it undertake in order to achieve its mission?

This is a fundamental step. An important aim of introducing new technology is to assist the achievement of the business mission of the department. All subsequent steps of the design task will be directed at the effective achievement of this mission.

B. Diagnosis of Needs.   The design group next discusses the following:

1. **Effectiveness needs. What critical activities related to mission achievement could it do better?** Here the design group focuses on those aspects of the department's work that most contribute to the business mission. Effectiveness is defined as "doing even better activities that are already being done well, and introducing new mission-related activities."
2. **Job satisfaction needs. What frustrations are preventing it from achieving job satisfaction and high morale?** For job satisfaction to be improved it has to be defined and measured. In ETHICS it is defined as: the fit between what individuals and groups are seeking from work (job expectations) and the reality of what they are receiving (job experience).

The definition of job satisfaction as a 'fit' means that the good and bad fits can be measured. The variables measured by ETHICS are:

Knowledge needs: the use and development of skills and knowledge.

Psychological needs: based on the Herzberg motivators (responsibility, recognition, status, sense of achievement, advancement).

Efficiency and effectiveness needs: the provision of required resources (training, information, supervisory assistance, acceptable controls).

Job design needs: work interest, use of skill, variety, discretion, autonomy, use of judgment and opportunity to take decisions.

The ETHICS method has its own questionnaire for eliciting this information (Mumford, 1983a).

3. Efficiency problems. What problems are hindering its ability to achieve its mission?

Variance analysis, a technique developed and used by the international group that operationalized the sociotechnical approach, is used here (Trist, 1981). A variance is defined as: a tendency for a system, or part of it, to deviate from some desired, or expected, norm or standard. The design group members therefore examine their own jobs and the work of the department as a whole to identify those parts where problems tend to occur. These are listed. The design group looks for two kinds of variance.

1. **Key variances.** These are deep seated problems which may be impossible to remove but which can be better controlled.

2. **Operation variances.** These are problems which arise through poor work procedures, poor organization, and shortage of information. Many of these can be removed altogether when the work system is redesigned.

The last part of the diagnosis requires an examination of

4. Future change needs. What changes are likely to take place in the future that will affect the ability to achieve the business mission?

Most systems today have a limited life span and have to be flexible enough to accommodate changes in the internal and external environment. It is important to consider these changes. All this diagnostic data is carefully recorded and transferred to a medium that makes it visible to all the group—for example, flip charts, transparencies, magnetic boards.

C. Information Requirements. The design group keeps the diagnosis of needs clearly in view as it answers the question:

What information will contribute most to the achievement of the work mission and the key tasks associated with this?

This information can be provided either by a computer or in other ways.

The design group is asked to identify the information **essential** to the following: "becoming more effective in achieving its mission," "obtaining more job satisfaction," "solving its major efficiency problems" and "managing change more easily."

This analysis of information needs is done twice more. Once to identify information that is *highly desirable* but not essential; and once to do the same for information that is *useful* but could be done without. The specification of **essential** information needs will provide a guide to what a computer-based information system must provide if it is to be successful. Information that is *highly desirable* and *useful* may be provided by a computer but can also be improved through better verbal and written communication. Clear objectives for the new system should now be agreed and discussed with the project steering group and the constituents of the design group members.

D. Departmental and Job Redesign.   Lastly, the design group will consider how the department as a whole, and individual and group work activities, can be redesigned to assist the achievement of the business mission, increase user job satisfaction, and make the best use of the technology. Sociotechnical design principles are often used at this stage. These are set out in the Appendix to this chapter.

ETHICS aims to give users some control over the system development process; an understanding of how the system will work, once implemented, and a feeling of system ownership.

THE ORIGIN OF ETHICS

ETHICS came about because of an experience I had 20 years ago when assisting a group of computer technologists to design and implement a new system. The systems analysts in an asbestos cement company in Manchester, England, were anxious to change the firm's sales office from a batch to a terminal based system for customer accounts. They approached me for help saying that they wanted to associate good organizational and job design with the new technical system. I did a survey of job satisfaction in the sales office and discussed the results of this with all the clerks, bringing them together in small groups. At these meetings a large number of organizational problems emerged and I suggested to the clerks that they should think about how these might be solved.

I then forgot about this request and fed-back the results of the survey to the members of the technical design group. They subsequently designed what they thought was an excellent sociotechnical system, called a meeting of all the clerks, described their proposed system, and sat back and waited for the applause. To their astonishment there was silence. Then one of the senior clerks stood up and said politely: "Thank you for your presentation, your ideas are quite good, but while you have been designing a new work structure for our office we have been doing the same thing, and this is how we should like to be organized." He then produced an excellent blueprint for a work structure that solved most of the office's efficiency and job satisfaction problems.

The systems analysts recognized the quality of the clerks' suggestions and it was the clerks' organizational solution that was used when the on-line system was implemented. It proved a great success. I personally learned a very important lesson from this experience—one that I have tried to apply ever since. This is never to underestimate a group's abilities. People at any level in a company, if given the opportunity and some help, can successfully play a major role in designing their own work systems.

This experience made me think very hard about how I could help other groups to participate in systems design and ETHICS gradually developed from my work with user design groups in other companies and situations.

Some Requirements of Successful Participation

A vital requirement is that users must want to make a major contribution to the design process and believe that their views will be listened to. Top management must be made aware of what is involved and promise their support, interest, and assistance. Similarly, the trade unions must not be opposed to participation and must believe that it is a genuine attempt to share decision taking and not a trick to make employees accept change that is not in their interest. A *facilitator* must also be found. Someone who will assume the role of guide and helper and assist a user design group to move purposefully along the road leading to a successful system. The facilitator should have good social skills, understand group processes and have experience of working with groups, be familiar with the ETHICS design method, and have a good knowledge of organizational and job design.

The person chosen as facilitator must be free from vested interest and so should not be a member of the user department or a systems analyst involved in the project. The facilitator's role will be to provide advice and guidance to the participants in the design process; help resolve any conflicts that arise and maintain the user group's interest and enthusiasm during the period that design is taking place.

Once the facilitator is recruited, the participative group structure needs to be created. Three decisions have to be taken at this stage. (a) What is the best kind of structure to adopt? (b) How shall group members be selected? (c) Who needs to be a member of any groups that are formed? The structure I favor is a steering committee and one or more design groups. The steering committee should consist of representatives of top management who have an interest in the system. These will include the senior managers of the user area, the head of management services, a senior personnel manager and, where appropriate, a senior trade union official. The role of the steering committee is to set the framework and guidelines for the project, to provide information on company policies and objectives, and to give support, guidance, and help throughout the period that the project takes place. The design group should represent all user interests. Its members will consist of staff of different grades and from different functions. A balance of age and sex is also important.

AN EXAMPLE OF USER PARTICIPATION IN DESIGN[1]

My two most recent projects have been assisting the Digital Equipment Corporation's U.S. sales force to design the software for an expert system and helping the staff of a number of British hospitals to define their information needs prior to the purchase of software. The first of these projects was published in 1989 by Wiley. The book is called XSEL's Progress: The continuing journey of an expert system and the authors are Enid Mumford and Bruce MacDonald. The second has not been published and is described here.

The British National Health Service is introducing comprehensive computer-based information systems. Its objective is to improve the management of resources by securing better information on how resources are used and how use affects cost. The British Government believes that improved management, assisted by more relevant, accurate, and timely information, will enhance the quality of patient care and enable resources to be used more effectively. The NHS is a complex organization. The needs of many different groups have to be identified and balanced for its effective operation, and inevitably there will be conflict between some groups. One new problem is the requirement for excellent cooperation between hospitals and regional health authorities in what is to become a buying and selling relationship. There is also the continuing requirement of cooperation between medical staff and administrators. All of this requires good management and clever computer systems. Computer software must be able to cope with considerable variety if it is to produce the information that different NHS groups regard as essential. The British Health Service is not only changing its organization and technology it is also changing its culture. Yet there is still much uncertainty concerning the advantages of the proposed changes, particularly the move to a pervasive computer technology.

Management at a small hospital in Manchester asked me if I would help it

---

[1]This project has produced a teaching package called the Trafford General Hospital Project. Package includes case study, workbook, and video. It can be obtained from the North West Regional Health Authority, Gateway House, Piccadilly South, Manchester, England.

implement its resource management program. Resource Management is the name given to the new market oriented organizational and technical change that is taking place. Linda Jones was the project manager and we agreed that a useful start would be to help the hospital nurses identify their information needs. This would enable relevant software to be acquired by the hospital for use in the wards. Nancy Brown, the senior nurse, gave an enthusiastic response to this suggestion and said she would like to play a leading role in the project.

Linda, Nancy, and I agreed that a small, representative group of nurses should become the user design group. This would include sisters and charge nurses, staff nurses, and enrolled nurses. These are the three nursing levels in British hospitals. We would share the facilitator role between us. ETHICS normally requires a series of about 10 meetings at 2 weekly intervals for the analysis to be completed. The hospital was unable to spare nurses for this length of time and we decided to compress the method into a 2½-day project. Two days would be devoted to the identification of information needs; the third half day would be for feedback and discussion of results. I would provide workbooks for the nurses to complete as they progressed through ETHICS and I suggested I make a video record of the discussions that took place.

On the first day Linda opened the meeting by describing the importance of improved resource management in running the hospital. Nancy then explained the nature of the task ahead. She said that it was essential to have a view on information requirements from all nursing levels, which was why representatives from the different nursing grades had been asked to become members of the design group. She suggested that each grade form a subgroup to work on ETHICS. I handed out the ETHICS workbooks and asked the nurses to fill these in as the different issues were discussed. I also asked permission to record their discussions on videotape, explaining that this could assist groups in other hospitals. Linda, Nancy, and I each sat with one of the subgroups to help them make progress.

After much discussion, each subgroup completed the first part of ETHICS and produced a mission statement and a set of key tasks and objectives. These were written down in their workbooks and on flip charts, and a member from each group explained how she and her colleagues had reached their conclusions. Ideas were discussed by everyone present and recorded on video. The flip charts were then collected by the facilitators and placed aside for future reference.

The nurses next moved onto the "diagnosis of needs" and described their nursing activities in more detail. They discussed and documented the following:

1. The tasks they had to perform on a regular basis.
2. What they spent most time on.
3. Their most frequent and most serious work problems.
4. Those aspects of work that required coordination with other groups.

5. New procedures they, or their wards, were introducing.

6. How they, and others, controlled their work. What kinds of targets were set and how these were monitored.

During the subgroup discussions ideas were noted in the workbooks and each group used one book as a master to record its final decisions before these were presented to all the members of the design group. Presentations were always followed by a general discussion of the points made and by questions to the presenter and her subgroup. At the end of the afternoon each of the design group members was asked to complete the job satisfaction questionnaire in the workbook.

On Day 2 of the project Nancy began the discussions by providing a verbal and written summary of the work of the previous day. She had brought together the ideas of the three subgroups and prepared a document that she hoped represented the views of the design group as a whole. The first part of this document set out a mission statement, followed by a summary of key tasks and key objectives. These are shown next.

NURSES MISSION STATEMENT

To be part of a multidisciplinary team providing a high quality of care to patients and their families within the resources available, and in so doing, at all times, to promote the better health, safety and well being of all concerned.

At all times to provide a professional environment which is conducive to the training and development of all staff, and which offers a controlled environment for learner nurses.

To provide a stable and rewarding atmosphere in which to work.

Key Objectives in the Attainment of This Mission are

To set achievable objectives for high quality care and to measure outcomes.

To provide a good learning environment which would assist staff development and high morale.

To provide a happy, relaxed atmosphere for patients, carers and families.

To make the most effective use of resources within budgetary constraints.

To be instrumental in assisting future developments and change in the Health Service.

Key Tasks are

Active participation in standard setting and the measurement of results.

An increase in understanding and self-awareness through attendance at courses and workshops on current issues and future developments in nursing policies and procedures, the process of nursing and models of nursing care.

Liaison with specialist groups and departments on specific issues such as ward budgeting and health promotion.

The second part of the document showed the tasks the nurses regarded as most important, what they spent most time on, their most serious problems, and their views on work coordination and control.

After discussing and approving this summary of their views the nurses began their next task, which was to answer the question "Where would better information help nurses most?" Deciding on this was to prove more difficult than their discussions of the previous day. They reexamined their mission, key tasks, and lists of nursing activities and considered what kind of information would be of most help to each of these. They prioritized their needs into *essential, highly desirable,* and *useful* information. The project was brought to a close in the late afternoon. We asked the nurses if they were now clearer about their information needs and the answer was a definite *yes.*

The nurses met Nancy for a final feedback session the following week. She had now prepared a second summary document, which set out their information needs in detail. The nurses approved this document and decided that a questionnaire should now be circulated to other nurses in the hospital. This would check whether all nurses agreed with the design group's definition of needs. If this proved to be the case the specification of information needs would be matched against available software. It would also be used as a guide for improving information processes through an enhancement of verbal and written communication.

Linda, Nancy, and I later evaluated the success of the project, agreeing that it had achieved our objectives. It had made the nurses more aware of the importance of good information and helped them to state where this was needed and how it could be provided. It had also given the nurses a degree of control over any new computer system that was introduced. They had influenced the choice of system. In the future they could monitor its performance and suggest how it might be improved. The project also helped the nurses to recognize that computers were not the only means for improving information. Improvement could also be secured through better verbal and written processes and procedures. The important thing was to recognize where and when problems existed and be prepared to work out solutions to these problems. The hospital now decided to extend the ETHICS approach to other groups.

An Evaluation of ETHICS

ETHICS has now been used in many companies by myself and others. Results have been proven positive. Acceptable systems have been created that users like and identify with. ETHICS has its own built in evaluation mechanism. The tools used at the diagnostic stage of the method can be used again once the system is installed and operational. The following questions are now asked. Has the new system made the user area more effective in achieving its work mission and key tasks? Has job satisfaction been improved and frustrations removed? Has efficiency improved through the elimination of old variances without the introduction of new ones. Is the system able to adapt to change that is taking place in the company? The answers to these questions can be both quantitative and qualitative.

Because ETHICS is a general problem-solving approach I have found that it is relevant to most situations although it may require modification. It can be used as described in this chapter or it can be regarded as a bag of tools. It has proved useful when a firm's own systems specialists are designing a technical system and when a package solution is being selected. Situations where modification will be required are times when organizations are at a particular stage in the design process—as in the hospital case, with very large systems and with systems where the organizational effects are still unresearched. Very large systems will require complex structures for effective participation, with steering, planning, and design groups being established at different locations and organizational levels. The XSEL project was an example of this complexity. The participation of all user staff in design, while desirable, is probably only possible in small organizations or at the grass roots where the design of a small part of a big system is taking place.

Another problem in today's use of ETHICS is that organizational solutions may be difficult to determine because the impact of a new form of information technology is not known. Ideally, organizational solutions should be created to achieve a business mission, not to fit technology, but it is clear that developments such as networks can have a dramatic effect on work structures and relationships. One way of tackling this problem is organizational prototyping. Different organizational structures can be tested out in pilot situations until the one that best assists business objectives and fits new technology is identified.

ETHICS can also be used in the development of standard packages but here it will be used in a very general way. Members of the design group will not be the future users of the system, but representatives of these drawn from different industries and occupational groups.

ETHICS was developed to assist the future users of a technical system to influence its design and to take responsibility for the design of the work system which surrounds the technology. But its use is changing and groups very different from the clerks and shop floor workers it was designed for are now using it. Nevertheless, I believe it still has relevance for lower level groups. The ethical reason for its development was to give groups with little power more strength and influence. These groups still exist and I suspect ETHICS may be of value to them.

APPENDIX

Guidelines for Individual and Group Work Design

Here are some principles of job design. Each job should provide the following:

Job Design Principles

1. A good fit with the needs of the person doing it. It should not be so routine as to cause boredom, nor so demanding as to cause stress.
2. Work variety and the opportunity to use a number of different skills.
3. The opportunity to use judgment and make decisions.
4. The opportunity to do a complete job—see a set of tasks through from start to finish.
5. The opportunity to learn and go on learning.
6. A feeling that work is important and seen by others as important.

The best jobs are those that contain the mix of routine and challenge that fits the personality and knowledge of the job holder.

Organizational Design

Here are some principles of organizational design that may help you.

1. People work best in small groups of six or less.
2. Giving a group responsibility for a mix of activities rather than a single function increases work interest and motivation.
3. Let a group identify and correct its own mistakes, rather than have another group do it. This prevents mistakes being made.
4. Try and ensure that information goes directly to the group that has to act on it. This avoids delays in taking action.
5. Give each group clear work objectives but leave it to them to decide how to achieve these. This encourages responsibility and stimulated initiative.
6. Make sure each group knows what it is responsible for and which groups it needs to cooperate with.
7. Give each group some development opportunities. For example, introducing new methods of work or new activities.
8. Involve staff in deciding what changes to introduce.

9. Keep your organizational structure flexible so that it can be easily changed.

REFERENCES

French, J., Israel, J., & Aas, O. (1960). An experience in participation in a Norwegian factory. *Human Relations, 13*, 2–19.

Lucas, J. R. (1976). *Democracy and participation.* London: Penguin.

Maclaren, R., et al. (1991). *Systems design methods: The human dimension.* New York: Digital Press.

Mumford, E. (1983a). *Designing human systems.* Manchester, England: Manchester Business School.

Mumford, E. (1983b). *Designing secretaries.* Manchester, England: Manchester Business School.

Mumford, E., & MacDonald, W. B. (1989). *XSEL's Progress: The continuing journey of an expert system.* London: Wiley.

Trist, E. (1981). *The evolution of socio-technical systems.* Ontario Ministry of Labour.

13 Principles in Practice: Two Cases of Situated Participatory Design

Kari Thoresen
Norwegian Computing Center, Oslo

The aim of this chapter is twofold: to describe some of our experiences with participatory design and to reflect on the relationship between methods and their use.

Many systems developers tend to view methods as precise directions to be followed rather than guidelines to be adapted to the situation. When methods must be adjusted or cannot be applied, common reactions are often that the method is inadequate, or that failure occurred because the method was not followed closely enough. Although systems developers in their daily work often adapt methods according to their needs, such adaptations are usually talked about with an air of apology; people feel they did not do things right, and thus, adaptations are conceived of as faults or deviations. It takes a fair amount of project experience, reflection, and self-confidence to be able to distinguish between the method itself and the practical use of the method.

Systems developers' attitudes to the use of methods are related to questions of participatory design in several ways. When people are faced with new dimensions in their practical work, they tend to deal with them in the same way as they deal with other aspects of their work. Their preconceptions of what systems development is about may shape the way they understand what participatory design is about. If they look upon methods as rules to be followed, it is likely that much effort will go into formulating rules for participation too, and less effort into finding out how to make participation successful. Also, if the rules are used ritualistically, the mutual benefits from participation may be reduced. This will undoubtedly confirm the belief of many systems developers that user participation is merely a nuisance and should be kept to a minimum.

Understanding systems developers' attitudes is important for us as researchers in the field. The action-oriented Norwegian research tradition, which is the basis for both projects described in this chapter, advocates working closely with both systems developers and users. It is vital for us to be able to introduce and discuss such topics in ways that both systems developers and users conceive of as useful. Our credibility as researchers in cooperative projects is partly founded in our understanding of not only the users' world, but also the systems developers'.

Focus.   We focus on evaluation activities in two different system development projects within the framework of participatory design. The first project, the Case Handling Project, developed a system to support case handling in Town Planning Departments in municipalities. The second project, the Hospital Project, developed a system to support nursing supervisors in a hospital in their personnel administration. Both projects used the same approach, i.e., evolutionary prototyping (see section 3), but were carried out rather differently due to the different settings. I illustrate how the settings impacted the system development process, with emphasis on the evaluation activities.

THE TWO CASES: SOME BACKGROUND INFORMATION

The Case Handling Project

The Case Handling Project took place in three Norwegian municipalities 1984–1987. It was a cooperation between a software manufacturer (the Vendor), the Town Planning Department in three municipalities, and the Norwegian Computing Center. We concentrate here on one of the municipalities, Farmland.

The project was initiated in a period of Norwegian political life where the expansion in the public sector was coming to a halt. The growth in employees in the public sector had been large, and there were signs of a political debate that would not allow further growth in the sector. The municipalities needed ways of signaling to the politicians and the public that they were willing to modernize and change their organization. Information technology was one of the answers.

Project Goals and Organization.   The common goal for the participants was to develop computer support for processing building permits. However, in a project with participants representing different contexts, there is bound to be different and, to some extent, conflicting goals in addition to the common one. The organizational differences between a municipal department, a private company like the Vendor, and a research institute, reflect their different aims and goals, values, work organization, environment, etc. Their attitudes and actions should be seen as reflecting these different contexts.

The Vendor's primary goal was to develop a new product to strengthen his position in the market; the municipalities wanted the system to contribute to simpler and more effective processing of building permits. As researchers, our interests were mostly on the methodological side. We wanted to try out evolutionary prototyping as a means of product development and to address the question of whether prototyping could be used to design one common system suited to many different municipalities. An important goal for us, as well as for the municipalities, was to try out ways to organize the development process so as to help the municipalities become less dependent on the Vendor.

The Vendor was in charge of the project. He provided the system developer, and a project manager who left his job with the Vendor towards the end of the first year of the project. Project management was then left (unofficially) to the Computing Center. The Town Planning Department participated with a local project group of 4–5 potential end-users, and the Norwegian Computing Center with two researchers.

This was the point of departure when the project started. Section 4 in this chapter describes some aspects of the process of development and evaluation. A more detailed description is given in Pape and Thoresen (1992).

The Hospital Project

The Hospital Project was a spin-off from a large work environment project carried out by the Work Research Institute in the same hospital. Early in the 1980s some of the surgical departments at the hospital were reorganized into one surgical unit. They had been independent units, but now they became sections within the surgical unit and shared admissions and an operating subunit. The reorganization was instigated in response to the economic crisis in the health sector, with the intention of reducing expenses by sharing resources.

Reorganization took place quickly, which held consequences for planning and procedures. The new organization implemented more standardized procedures than when the units were independent. The need for work procedures adapted to the conditions of the different sections was not taken seriously enough. Planning procedures may serve as an example. Sections with only 20% patients in need of immediate surgery can use other planning procedures than sections with 80% patients in need of immediate surgery.

The need for planning information for the whole unit, not only each section, increased. At the same time, work pressure was so high that it was hard for the nursing supervisors to deliver necessary data in time. The unit supervisor had a tough time getting the information she needed from the nursing supervisors.

This was not the only problem related to the reorganization. The Work Research Institute had been engaged at the Hospital for some time, carrying out a participatory, action-oriented work environment project. One researcher suggested that the information-processing problems mentioned earlier might be improved by information technology. The Norwegian Computing Center was involved, and when the opportunity for an information technology project developed, the nursing supervisors were willing to participate. The Hospital project took place from 1987–1989 as a cooperation between the nursing supervisors, the Work Research Institute, and the Norwegian Computing Center.

Project Goals and Organization.   Nursing supervisors are responsible for nearly 500 nurses. Administering the nurses relative to the needs of the patients is a complicated piece of work. Supervisors wanted to use the project to improve their own coordination efforts. They needed to reduce their paper work and have a system that supported their local personnel administration.

Researchers wanted the project to contribute to the use of information technology systems supporting communication and cooperation in service-oriented work. This was done by using the project and the design process as an opportunity for training for several types of cooperation (Bermann & Thoresen, 1990):

- *Cooperative work:* by strengthening cooperation between the unit manager and nursing supervisors through project work.

- *Cooperative systems development:* by organizing the development process as a mutual learning process for the nurses and researchers involved in the project.

- *Computer support for cooperative work:* in the sense that the information technology system should enhance rather than hamper cooperation among the nursing supervisors.

Personnel administration for surgical wards must reflect the characteristics of the work situation. The setting is turbulent, replanning is constant, and immediate help is often needed. Thus, design concepts from other types of personnel administration should not be transferred uncritically to systems designed for administrating nurses.

The characteristics of service work can only be understood in relation to the customers, whether they are patients or hotel guests. A characteristic feature of patients is that they appear in many shapes and have many different needs and unpredictable demands. Thus, rule-based procedures may not be adequate as a design concept for service-oriented work. Another project goal was, therefore, to develop design concepts that would capture the particular characteristics of service work.

The final project goal was to test different techniques for learning during the design and development of an information technology system. The following story concentrates on this particular goal.

In addition to 6 nursing supervisors, each for one of the sections in the surgical unit, and their unit manager, the project team consisted of three researchers, one of them functioning as the system developer, and a graduate student, all participating on a part-time basis.

Similarities Between the Projects

Both cases used evolutionary prototyping. The rationale behind the approach was to use the potential inherent in the design and use of information technology systems as a catalyst for change. Thus the projects are based in the Norwegian research tradition of action oriented work life research.

The term *action research* is known mostly from social science research, but in Scandinavia action research has also been used for a number of projects related to information technology. In the 1970s several Norwegian trade unions cooperated with researchers to develop trade union based strategies for dealing with what was then seen as the coming wave of computerization (Keul, 1983).

A characteristic difference between action research and other research projects is the question of what is a result. While *results* from a research project are often descriptions, analyses, and contributions to theory of the discipline in question, *results* in action research also include the realization of some practical or political improvement in people's lives, for example, in their working conditions. Action research alternates between practical work in the field to support the desired changes, and systematic data collection and analysis of the practical work with the aim of improving the action.

The term *action oriented* for our two cases points to the relationship with the action research. The change ambition is illustrated in the Case Handling Project by the goal to enable the employees in the Town Planning Department to develop necessary knowledge to manage technological and organizational change on their own, without undue dependence on the Vendor. In the Hospital Project, the desired change is not only to provide the nursing supervisors with a software system, but to organize the process so that they would learn how to benefit from information technology, and to strengthen their ability to cooperate, both in projects and in their daily work.

Both projects were small, involving only one designer/programmer in addition to 2–3 researchers, all on a part-time basis, but with a considerable amount of user involvement. Both projects were multidisciplinary, in the sense that the researchers represented social science and computer science.

THE DEVELOPMENT APPROACH

Evolutionary prototyping denotes a development approach where information technology system and organization is developed simultaneously and integrated. Floyd (1984) defines evolutionary prototyping as a sequence of cycles: (re-)design, (re-)implementation, (re-)evaluation. The emphasis is on software constructed within a dynamic and changing environment. Instead of trying to capture a complete set of requirements in advance, the system is built so as to accommodate subsequent, even unpredictable changes. The approach relies on individual and organizational learning, by systematically collecting experiences with using the system in practical work.

A more precise term could have been *gradual, iterative development integrated with use.* For the sake of simplicity in our daily cooperation, we used the term *prototyping.* The term prototyping, nevertheless, may cause confusion when you talk with systems developers because they commonly associate prototyping with designing screen formats to be presented to the users for evaluation. In such cases, prototyping is mainly a technique for visualizing design suggestions in order to improve requirements specifications during early phases. On the contrary, the evolutionary prototyping in our cases was a somewhat different venture. In this chapter, however, we use the term *prototyping* to mean evolutionary prototyping.

Steps in Evolutionary Prototyping

This section describes the approach in general terms. In following sections we describe how the approach was adjusted differently to the different setting of the two cases with particular reference to the evaluation.

Figure 13.1 is a high-level illustration of the sequence of steps. The soft lines indicate that each step cannot be prescribed in detail; organizing and carrying out the practical activities should be adapted to the setting at hand. The starting point is thorough information and discussion about the evolutionary approach. Both systems developers and users need to understand the rationale behind the approach and how evolutionary prototyping differs from waterfall models. Training

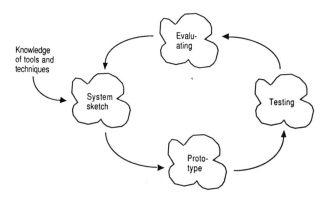

FIG. 13.1.   Sequence of steps in prototyping.

in the development tools to be used, for example, software and techniques for analysis and design, is also necessary for both users and system developers.

System Sketch.   The next step is making a system sketch, in cooperation with the users and the system developer(s). The term *sketch* is chosen to emphasize that it is preliminary, and to distinguish it from a specification. The syntax and semantics of the sketch are less important than its potential as a means of communication.

Designing and Constructing the Prototype.   The next step is to design and construct the prototype. In both cases the prototypes were working prototypes, not merely demos. They had a limited number of functions, but these functions were fully developed so that the prototype could be used in daily work. Having only a few of the desired functions is definitely inconvenient for the users. A prerequisite for using this approach is that the users taking part in the development are thoroughly informed and willing to take on this extra trouble.

Testing the Prototype.   The prototype should be tested out in the users' regular work. The testing in practical use is a critical point for several reasons. It takes time to build up an understanding of how to use the system. Without this understanding, the users' contribution to the evaluation may be superficial, bringing forth only the obvious flaws or the cosmetic parts. Particularly the relationship between work organization and system design is not easily discovered unless tested out in practical use. Such a relationship consists of several, often contradictory relations. They are often subtle, and they reveal themselves only gradually. Discovering the range of organizational opportunities and problems requires practical use of the system for extended periods.

Evaluating and Deciding Priorities.   The next step is the evaluation. The evaluation serves as a way of assessing whether the process and product requirements have been satisfactorily fulfilled so far. It also functions as a search process, and the information is fed back into the project to adjust and improve the process and the product at an early stage. It is important to note that the evaluation is concerned not only with the prototype, but with the changes in work routines, the training, the participation aspects, etc. Setting the agenda for an evaluation should be dealt with early in the project, to give the participants opportunities to prepare for an expanded set of topics to be discussed. The evaluation was carried out very differently in the two cases, nevertheless the underlying principles were to a large extent the same.

One part of the evaluation is to determine priorities for the next iteration. What is not conveyed in Fig. 13.1 is that this task relies upon a considerable degree of mutual trust and shared knowledge. You need to be both critical and self-critical; you may even need to throw away the whole thing and start again.

For an evaluation to be meaningful, the users must trust the professional competence and the seriousness of the systems developers and vice versa. If not, the whole basis for cooperation may deteriorate.

In the next iteration the sketch is further developed, adding new functions and refining the old ones. The new prototype version is (re)designed, (re)constructed, tested out in practice again, and so on. The iterations continue until some predefined limit is reached; a specific number of iterations, or when requirements are fulfilled, or until the users are satisfied, or, as often happens, until one runs out of money.

It is important to note that evolutionary prototyping as previously sketched must be embedded in a project plan and a project organization to secure appropriate control. The risks of anarchy are considerable unless care is taken to follow higher level goals and plans.

THE CASE HANDLING PROJECT: EVALUATING
BY SEARCH CONFERENCES

The requirement to develop a product to be marketed for 400 municipalities called for a special project and process design. The strategy chosen is illustrated in Fig. 13.2 and described in more detail in Pape and Thoresen (1987, 1992). Next, the main points in the development strategy are explained with reference to Fig. 13.2.

Development Strategy.   The development of the First prototype in Farmland started with the researchers carrying out a sociotechnical analysis of the Town Planning Department in Farmland; the work of the potential users, the laws and regulations to control building activity, and the relationship between the Town Planning Department and the environment. The analysis supplied data needed for the design of the First prototype and for the organizational implementation. The prototype was constructed and tested in production in Farmland. After the test period, the first evaluation was carried out, and the results of the evaluation were input to the next prototype and the organizational changes. The full sequence of steps was repeated twice, thus in Farmland there were in all three cycles of design, testing and evaluation.

During the development period for the First prototype, a representative from Coastland sat in on project meetings, to learn from the experiences in Farmland, and to identify the areas of system design where changes were needed for Coastland. A sociotechnical analysis was carried out in Coastland, and when the Second prototype was ready for use, it was to be installed in Coastland. The experiences from Farmland was used to design a plan for training and organizational implementation in Coastland. Again, the same sequence of steps was

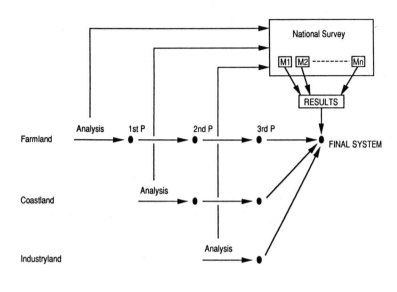

P = Prototypes
M = Municipalities

FIG. 13.2.   Development strategy. Redrawn from Pape and Thoresen (1987). Reproduced by permission.

carried out as in Farmland, but since Coastland was *one cycle behind* Farmland, only one iteration was possible.

After evaluating the Second prototype in both municipalities, the experiences to date were collected and fed back into the last version of the prototype, which in turn was tested out in a third municipality, Industryland. This municipality had no opportunity to influence the design before the system was installed, and thus functioned as a test of the quality of the design.

The national survey (see Fig. 13.2) was a questionnaire survey of 25% of the Norwegian municipalities. It was designed to identify the spectrum of variations relevant to the design and use of such systems, and also gave us some indications of whether Farmland, Coastland, and Industryland were representative for the target population.

Project Setting in Farmland.   Farmland is situated 500 miles away from the Computing Center. Whereas the Vendor and the system developer had easy access to the Town Planning Department, the researchers did not because of the distance. It was not possible to drop in now and then to see how the municipality progressed. On the other hand, it was vital for us as researchers to be able to follow up the local processes, and if necessary, take action. The evaluations represented one of our few opportunities to see for ourselves how the local processes developed. First, however, we review some information about the project progress in Farmland before the evaluation.

The secretary, the clerk, the case handler, and the department engineer, all potential end-users, were to be active participants in the project. The department engineer was also the local project manager. They were to take part in all project meetings and have access to all project documents, as prescribed by Norwegian law.

It soon turned out that the employees did not attend project meetings regularly, except for the engineer. We reminded the engineer again and again that the other employees were wanted. Now and then, one or two of them turned up, polite and friendly, but they were obviously bored with the meetings. Compared with our previous experiences, the whole situation did not seem very participatory, and above all, much too peaceful. We expected conflicts and resistance, not a big, happy family.

Evaluating: For Different Purposes

When the time came for the first evaluation we did not know what to expect. Perhaps the seemingly quiet situation was some kind of a cover-up, controlled by the local project manager who had means to silence the employees? Frequent contacts with the employees would have given us the opportunity to see for ourselves what happened. This was not possible, however. The distance between the Computing Center and the municipality made it too costly to keep in close contact with the employees, we had to trust the reports from the system developer and the department engineer. They were both positive. According to them, several system sketches were made, and it was decided which sketch was to be the basis for the first prototype which was implemented and put into productive use. We felt that we could not press for participation from the other employees more than we already had. That would indicate lack of trust towards the local project manager and the system developer. Therefore, the evaluation became more important to us than we had originally anticipated. It was intended partly as a measurement device for the process and prototype so far, and partly as a search process to discover new opportunities and problems. Now it also became our opportunity to find out what was *really going on,* and whether further action was needed.

The Principles of a Search Conference

The evaluation was carried out according to the principles of a Search Conference, a methodology developed by the Work Research Institute in Oslo in the 1960s, and rooted in the Norwegian tradition of cooperative, action oriented work life research.

A Search Conference is a way of organizing a company-based conference to establish a common starting point for change, and to do it in a democratic way. The idea behind the conference is to bring together groups of employees who normally do not meet for such discussions. The aims are to establish a higher degree of shared goals and more flexible channels of communication than permitted in bureaucratic structures.

The conference alternates between group work and plenary discussions, and the composition of the groups changes between horizontal groups (e.g., all the secretaries), and functional core groups (e.g., the employees belonging to one particular department, from top to bottom of the ladder). The methodology relies heavily on bringing forward different perspectives and legitimating these through discussions. "Rules of the game" are established to provide for such discussions. Rules may concern length of discussions, how to elect chairpersons, the documents to be produced from the discussions, etc. It is emphasized that the groups need not reach agreement, and each group decides for itself how different opinions should be expressed in plenum.

The conference is built around the core concept of democratic dialogue (Gustavsen, 1985). The criteria for such dialogues included the right to participate for all employees who will be influenced by the project, the duty of every participant not only to bring forward his or her own opinion, but also see to it that other participants' opinions are brought forward. Work experience is the knowledge basis for participation, and all arguments concerning the question discussed are legitimate. An argument may be rejected, but not until after it has been discussed.

The researchers function as a secretariate, organizing the conference but not interfering with the discussions unless it is necessary (e.g., when the rules are violated). They also sum up the results every day, and write the final report, in cooperation with the participants.

Trying out the Search Conference

Because the Search Conference is designed for larger organizations, we could only use the principles, not the more detailed guidelines (Palshaugen, 1986). It was organized as a 2-day conference in the Town Planning Department, with the employees, the system developer, and several others in the department; among others, the Head of Department who was not a regular participant in the project. Each of the conference participants was interviewed, and they all met for group discussions the second day. The secretary and the case handler demonstrated how they used the prototype and what they wanted to change.

Looking Back

Some experiences from this evaluation seem worth mentioning. First of all, the *calm waters* were real, not some facade put up by the department engineer and the system developer. There had been a number of problems with the Vendor and the equipment; delivery times, operational stability, etc., but very few with the employees' cooperation with the local manager and the system developer. They had been active in dealing with the problems themselves, this was a part of their work culture. This served to remind us of the fact that participatory design occurs not primarily in project meetings, but during the day-to-day cooperation between the employees and the system developer.

Even if the local process had proceeded smoothly without any interventions from us, several topics emerged during the Conference which brought some movement into the project and into the Town Planning Department. An example may be the secretary's suggestion that they could save time for the customers if she was authorized to sign the building permits when the case handler was not present. It had been discussed internally for some time, but was taken more seriously now than earlier. Suggestions from the employees may often be taken more seriously when they are sanctioned by a research project, and this may be part of the story in our case too. It was clear, however, that the Conference contributed in two ways: It helped to give people a richer picture of their own work place, an understanding of their colleagues' perspectives, and it provided an opportunity to formalize decisions that had been informally discussed for some time.

The secretariate's position was important. From the interviews we learned about possible topics to be discussed in common. Not all topics could be discussed publicly, however. We also learned which topics should not be talked about. Functioning as a secretariate provides you with some power to set the agenda and also to direct the group discussions. Because we interviewed all the participants it was possible to minimize attempts from single actors to form alliances.

On the other hand, this point reflects parts of the criticism voiced by opponents to the Norwegian tradition of cooperation. The critics point to the lack of class perspective underlying the conference philosophy, and argues that the emphasis on common goals may reduce the legitimacy of interest-based arguments. It is only fair to say that the critics do have a point.

THE HOSPITAL PROJECT: INTEGRATED EVALUATION

Project Setting at the Hospital.   The conditions for project work were much less favorable in the hospital than in the Town Planning Department. We were accustomed to project meetings in organizations with traditions for project work, where people knew the rules of the game and behaved accordingly. This was not the case in the hospital. Considerable effort was required to understand each others' limitations and abilities, and it took us some time to master the situation.

Throughout the project period at the hospital it was difficult to carry out meetings. The nursing supervisors were always running to and fro for something else, they all wore pagers, which at times made it difficult to keep the continuity. At the most intense times, somebody was talking on the telephone at the same time as another person's pager was sounding, and a third person was on her way to deal with some emergency, while the rest of us tried to carry out a decent project meeting. The work overload of the nursing supervisors called for ingenuity in organizing and carrying out the project so as not to increase their work load further and unnecessarily.

There was also a lot of complaining—about the hospital, the economy, the administrative routines, each other, us, the project, the lack of computer terminals, etc. The nursing supervisors certainly had good reasons to complain, but it made cooperation stressful in the beginning. Other researchers have experienced the same. It eased the situation when it became clear that this was not directed at us in particular but was part of the hospital culture.

After some time we realized that what we conceived of as unrest and disturbances were closely related to the characteristics of the hospital as a work place, in particular the surgical unit with its turbulent work situation. The nursing supervisors became carriers of these characteristics.

Integrating the Evaluation

In such a setting, organizing a search conference was doomed to failure. Having everybody available for 2 entire days without interruption was unthinkable. Our solution was to integrate the evaluation into the regular project work and make it an ongoing activity.

Joint Meetings.   All project meetings were joint meetings between the researchers, including the system developer, and the nursing supervisors. The joint meetings contributed to a mutual learning process. They were not only an advantage for the nursing supervisors, but also for the researchers who learned more about the supervisors' work situation. Bringing to the surface reactions and feedback, both positive and negative, helped speed up the learning process. Face-to-face contact makes learning easier than communicating by documents, like requirements specification and minutes for meetings.

Frequent Meetings.   Our goal was to have a joint meeting every second week, which was almost fulfilled. Nevertheless, these were not too popular with the nursing supervisors, and there was some mumbling about "all those meetings." The mumbling was not to be neglected, however. When the complaints became loud, we reduced the length of the meetings, but not their frequency. If we did not have frequent meetings, there would have been far more misunderstandings to clear up and far more system errors discovered too late.

Another important aspect of the project meetings was their function as a meeting place for discussing work-related issues outside the scope of the project.

These discussions were often triggered spontaneously by some point related to the prototype, but often ended up addressing other, seemingly irrelevant, aspects of work. In the long run it became clear that what seemed irrelevant at the time, could be of vital importance for the coordination problems of the nursing supervisors. Thus the project meetings supported mutual learning among the nursing supervisors about each other's work, what they did differently, how and why. Again it helped to legitimize different ways of looking at their work.

Evaluation: Individual and Joint

The infrastructure of the evaluation was the project meetings. This does not mean that the evaluation was carried out in meetings only, the nursing supervisors' own use of the prototype was the main basis. However, the meetings provided a common ground, which was used to make the evaluation not only an individual activity, but a joint effort.

The Tutor. The graduate student played an important part in the evaluation. She functioned partly as a tutor, and partly as a collector of information for the joint evaluation. One of the reasons for introducing the tutor was to solve the problems of getting all the supervisors to use the prototype. The differences in how it was used were wide. Some used it every day and experimented on their own; others were content just to log in and out.

These differences might jeopardize the joint evaluation. Those with little experience in using the prototype would not be able to contribute in the same way as the skilled ones. The knowledge basis necessary for evaluation would simply not be present. We might have chosen to let those lagging behind continue to lag behind, but taking into consideration the project goal of improving cooperation and coordination, we could not risk losing them.

The tutor made appointments with each of the nursing supervisors to help them in their office, a few hours every second week, in teaching new functions and helping solve problems of use. Similar tutor activities have been described from other projects (MacLean, Carter, Lovstrand, & Moran, 1990).

The fact that the tutor had personal contact with each of the nursing supervisors proved advantageous. She was often present while the nursing supervisors used the prototype in their own office, thus presenting her with the opportunity to hear them "talking-in-work," not only "talking-about-work." The nursing supervisors also felt more committed to practice on their own. It was too easy to excuse themselves in project meetings, saying they had not had the time to practice; it was not that easy when they had an appointment with somebody who was there to help them. The tutor not only helped; she also took notes about the nursing supervisors' opinions of the prototype—what to change and how. The notes were presented regularly at project meetings and were part of the basis for evaluation and redesign.

There were several factors that helped to overcome the problems of some nursing supervisors lagging behind. The tutor was one, another was the gradual improvement of the prototype, and yet another factor was the hospital's decision to let one of the nursing supervisors function as a full-time development nurse. She could spend more time practicing and experimenting with new ways of using the prototype; she became an in-house tutor and a valuable support for both her supervisor colleagues and for the system developer.

Logbooks and Demos.   Logbooks and diaries written by the participants are a familiar way of attempting to capture the day-to-day experiences. The trouble with them is that they often do not work. Logbooks were tried out in the Case Handling Project with little success because they were used too sporadically. This will probably be the case as long as the benefits are, or appear to be, mainly on the researchers' side.

With the nursing supervisors, logging results were different for several reasons: The tutor used the logbooks as sources for her notetaking, and thus, the nursing supervisors knew that their comments were not wasted; they would be part of a discussion in a project meeting. Their comments became legitimated through the process of being recorded and presented later for discussion.

Demonstrating how to use the prototype was another technique tested. In the Case Handling Project such demos did not function well for evaluation purposes. They did not capture the many possible varieties of use, in particular, since the formal evaluation activities during the search conference lasted only a couple of days. With the nursing supervisors, demonstrations of use functioned quite well. We managed to establish a wider group of critics than the project participants. There were observers from other parts of the hospital who came back to the meetings many times, with their own ideas about computers in nursing. In this way a continuity of criticism and discussion with the world outside the project was maintained. The continuity was made possible because the evaluation was integrated into the regular meetings over an extended period of time.

A special type of demo turned out to be valuable in developing and encouraging the use of the prototype. Each nursing supervisor took her turn demonstrating to the other participants how she used the prototype. These sessions contributed greatly to spreading ideas for using the prototype and illustrated creative (and sometimes disruptive) new ways of working around the system, in the sense of intentionally using computing in ways for which it was not designed (Gasser, 1986). These *working arounds* in turn gave ideas for improvements in the prototype and increased mutual learning.

It took some time, however, before such sessions were accepted. Some of the nursing supervisors tended to interpret the demos as directives and prescriptions. The reaction was understandable due to their long experience with directives as a means of effecting changes. When it became clear that these sessions were to be seen as sources for sharing ideas rather than giving orders, they became viewed as valuable by both the researchers and nursing supervisors.

Some Differences Between the Projects

There was no vendor involved in the Hospital project, which gave us considerably more freedom in carrying out the project. On the other hand the lack of a vendor made it more difficult to secure continuity regarding the life of the prototype. There was a risk of *losing* the prototype in the sense that the hospital could or would not provide technical support and further development after the project.

The nursing supervisors' willingness to participate depended very much on our willingness to take this problem seriously. This meant a lot of work both for us and the nursing supervisors to secure the future for the prototype. It was necessary to work closely with the hospital's systems department, to choose equipment and development tools compatible with the (at that time unofficial) information technology strategy of the hospital, to increase the visibility of the project, and to obtain and renew management support. Some may call such activities *selling the project*. The language of sales and marketing, however, give only a shallow understanding of the type of work involved in such activities.

CONCLUSION

In both reviewed cases methodological ideas have been applied and adjusted to different contexts. Some might look at each of these approaches as system development methods. It seems more appropriate to view system development as an ongoing process where methdological ideas are applied and adapted to the setting at hand. One should draw on the principles and underlying ideas that are embedded in the methods, but parts of the setting are always unique and should be allowed to impact on how the methods are used. In this sense, qualified use of methods may be looked upon as situated actions (Suchman, 1987).

It helps to have a repertoire of techniques to draw from, and it helps to have systematic and empirically based knowledge of the practical use of such techniques, not only vague ideas of what worked and what did not work. But in addition to facts, like techniques and empirical data, there will always be an element of surprise, of art and creativity. Not everybody will like that—neither the unions who often go by the rules and regulations, nor the employers who often want to know exactly what they are buying. Participatory design does have a dimension of exploring that is part of the fun.

ACKNOWLEDGMENTS

Thanks to Tamar Bermann who initiated and participated in the Hospital Project, for the cooperation, the inspiration and the wealth of knowledge she shared with us during these years before her death in September 1990. Thanks also to Tom

Chr. Pape for his contributions to the Case Handling Project and to this paper, and to the anonymous reviewer.

The Norwegian Research Council for Science and Technology financed the researchers in both projects.

REFERENCES

Bermann, T., & Thoresen, K. (1990). Can networks make an organization? In G. Bjerknes, B. Dahlbom, L. Mathiassen, M. Nurminen, J. Stage, K. Thoresen, P. Vendelbo, & I. Aaen (Eds.), *Organizational competence in system development* (pp. 53–65). Stockholm: Studentlitteratur/Chartwell, Bratt Ltd.

Floyd, C. (1984). A systematic look at prototyping. In R. Budde, K. Kuhlenkamp, L. Mathiassen, & H. Züllighoven (Eds.), *Approaches to prototyping* (pp. 1–18). Berlin: Springer-Verlag

Gasser, L. (1986). The integration of computing and routine work. *ACM Transactions on Office Information Systems, 4*(3), pp. 205–225.

Gustavsen, B. (1985). Workplace reform & democratic dialogue. *Economic and Industrial Democracy, 6,* pp. 461–479.

Keul, V. (1983). The trade union movement: Research and data technology. In E. Fossum (Ed.), *Computerization of working life* (pp. 21–61). London: Ellis Horwood Ltd.

MacLean, A., Carter, K., Lovstrand, L., & Moran, T. (1990). User-tailorable systems: Pressing the issues with buttons. *Proceedings of CHI'90: Conference on Human Factors in Computing Systems* (pp. 175–191), Seattle, WA. New York: Association for Computing Machinery.

Pape, T. C., & Thoresen, K. (1987). Development of common systems by prototyping. In G. Bjerknes, P. Ehn, & M. Kyng (Eds.), *Computers and democracy: A Scandinavian challenge* (pp. 297–311). Aldershot, UK: Avebury.

Pape, T. C., & Thoresen, K. (1992). Evolutionary prototyping in a change perspective. *Information: Technology and People.*

Palshaugen, O. (1986). *Means of designing a starting conference.* (Work Research Institute, AI Doc. 28/86). Oslo: Work Research Institute.

Suchman, L. (1987). *Plans and situated actions.* Cambridge, UK: Cambridge University Press.

14 Small Changes: Starting a Participatory Design Process by Giving Participants a Voice

Joan Greenbaum
La Guardia College, City University of New York

Kim Halskov Madsen
Aarhus University, Denmark

Those of us in the computer field often tend to see our world as one where things happen rather fast. In the area of computer system development the '80s was a decade where an overwhelming number of new system methodologies were introduced, and where the overall perspective on system development shifted to include varying degrees of user involvement. Each time a change was trumpeted in a journal or at a conference there was often a chorus of researchers ready to announce the change, as if the world had in fact switched to the new idea. But in the day-to-day practice of developing computer systems, as in most practices, changes are slow, and, in fact quite small.

In this chapter we don't bemoan the fact that practice changes in small stages, but rather we celebrate some of the changes taking place and point to areas where further small, inexpensive, and, rather uncomplicated changes can continue to take place. The area we focus on is that fuzzy field covering the interactions between system developers and end-users. We avoid the overworked and often misused term *end-user* and instead concentrate on the people within an organization who can be encouraged and allowed to participate in making decisions about computer support in their workplace.

There has been some discussion recently, for instance at the Participatory Design Conference in Seattle in 1990, concerning the applicability of Scandinavian approaches in the United States. Although we can't begin to give a definitive answer to that question, this chapter tells the story of a project in which we applied some techniques from a Cooperative Design approach (Greenbaum & Kyng, 1991) in an American setting. In particular, we used a variety of workshop techniques to help staff members in an organization better articulate their needs concerning desktop hardware and software. Our example is illustrative of ways

289

that decisions concerning computer support can be done cooperatively. As in all stories there were some successes and a few downbeat moments, but we are convinced that the road to small changes is paved with these little incidents.

The specific case discussed here concerns a project that was part of the Management Information System (MIS) department's planning activities at the headquarters of a large American nonprofit organization. As independent consultants we were asked to involve staff members in the ongoing MIS planning activities. The goals of the project were: (a) to enable participants to better articulate their current work activities and formulate their future expectations concerning computer support for these activities, (b) to create an ongoing learning environment where participants could learn from each other and pass their learning on to others, and (c) to provide MIS management with information for directional support concerning desktop computer software, specifically concerning word processing, spreadsheets, and databases.

APPROACH TAKEN

The project, called the Participatory Design Project or PDP, was initiated by senior management's request for what they called directional support concerning decisions that they wanted to make for the organization's desktop computing needs. The MIS department had prepared a first phase study of organizational computing needs that outlined the possibility of using some kind of local area network (LAN) to link desktop computers. The plan also indicated that the organization would need to switch to some form of standardized hardware and software in order to foster communication among computers and let the office staff share both information and computer resources. The initial MIS plan included interviews with department directors. We were invited to submit a proposal that would suggest ways to find out if the original assumptions were valid. Our approach was to involve people who worked with computers on a day-to-day basis and, over a period of 2½ months (constrained, by other management decisions within the organization) we undertook the following:

- Project establishment
- Storytelling workshop
- Critique workshop
- Fantasy workshop
- Implementation workshop
- Vendor visits
- Reporting

In our role as consultants we served as facilitators for each of the workshops and helped plan the administrative details of getting participants involved in the project, as well as working closely with them in preparing our final report and coordinating their presentation to management. Additionally, we gathered technical details about LAN hardware and software, and information about specific desktop packages in order to better acquaint the participants with the range of technical options open to them.

Because the goals of the project were both to give management directional support and to create an ongoing learning environment for the office staff, we needed to coordinate technical and organizational information that would support the needs of both groups. As it turned out we were able to do this to the satisfaction of both sides. But this was probably due to the cooperative and helpful spirit in the workplace, and less to our prior understanding of how to handle the situation. The issue of double and sometimes conflicting objectives is not directly handled in either American or Scandinavian system development literature and clearly needs more discussion among those practicing design.

Project Establishment

In selecting the participants, department heads were asked to involve people who had the most day-to-day experience using existing software (Bødker, Greenbaum, & Kyng, 1991). Twelve staff members were selected to participate in the project. We asked the department heads to recommend people as that was the protocol in the organization; some people ended up volunteering with the approval of their managers. The 12 participants were from a range of job titles including editors, editorial assistants, billing clerks, accountants, and secretaries. The group was fairly homogeneous in that they were quite skilled and extremely motivated.

Senior management at the organization agreed that the participants could be released from their regular activities to take part in up to 6, half-day workshops. Although all of the staff were extremely busy, both management and the staff participants felt they would benefit from the workshop process. There were some negotiations and compromise over which departments and which workers should be included. This, of course, could be expected. As a compromise we ended up involving three staff members from four of the five departments, although our original suggestion was to do more in depth work with one department. We return to this issue later as the multidepartmental workshops turned out to be extremely fruitful.

Storytelling Workshop

For the first workshop the participants were asked to prepare two short oral stories about their worst and most successful use of computers in their daily work. At the first meeting the participants shared these stories in order to initiate discussions about future software needs. One of the stories told by a staff member, was about the day she ended up spending 2 hours trying to print out a letter. She went from one personal computer to another running into all sorts of problems (i.e., the lack of the right size disk drive, lack of memory, and incompatible printers).

The storytelling workshop was important in two respects. First, and that was our conscious intention, the workshop provided the participants with an opportunity to become familiar with the situations in each other's offices. Second, and this turned out to be even more important, the participants also came to realize that the daily problems they faced were not unique. During the evaluation of the project participants typically stated "I didn't know that other people here have the same problems that I have."

Critique Workshop

The following three workshops were organized accordingly to a Future Workshops approach, developed by Jungk and Müllert in Germany (Jungk & Müllert, 1987). This approach was originally used as a technique to enable citizen groups to take part in the decision process of public planning authorities such as town planning, and environmental issues. Specifically these types of workshops are organized to generate ideas for future activities and to initiate actions for implementing those ideas. Future Workshops are also designed to be conducted in a way that supports people who are not typically involved in politics or other kinds of decision making. Recently the Future Workshop approach has been shown to be useful as a way of involving users in systems design (Kensing & Madsen, 1991).

A Future Workshop is typically held in three phases: Critique, Fantasy, and Implementation. In the Critique phase the purpose is to give voice to current problems and structure them so that a constructive outcome can be expected. In the Fantasy phase the purpose is to generate visions about an ideal future. The Implementation phase proposes to evaluate the visions and to set up a plan of action. The approach seems to work best when each phase is facilitated by someone who is outside the process but who is familiar with the organizational issues. Additionally, it helps to arrange the workshops in a way that participants are in a comparatively *safe* situation within the organizational hierarchy so that they are not afraid to voice their problems.

A critique workshop generally starts by having people brainstorm about critical issues. Specific brainstorming guidelines for a future workshop include:

- speaking time is restricted to 30 seconds;
- each point of critique should be formulated as a short statement and written on a wall chart for all to see;
- there is no need for rational arguments; simply stating an idea is enough;

- the participants are not allowed to enter a discussion or argue against each others' points of critique.

These guidelines, while perhaps sounding rather rigid, turn out to help support participants who are sometimes afraid to speak or feel awkward phrasing their ideas. The Women's movement has often used similar guidelines, like "going around the room to hear from participants" in order to enable women to speak out (Bødker & Greenbaum, 1988).

The next step is to find three central themes from the critique and then group the critique statements according to the themes. Afterward, the participants are split into teams, each responsible for elaborating one of the themes during group discussions. At the end of the critique phase each team gives a short presentation of their elaborated critique. The presentation can be in any form found appropriate by the team, for instance a small speech, a poster, a picture, a caricature, or a small play. People are encouraged to be as specific as possible by pointing out concrete situations where problems have arisen. At this point in time, participants are encouraged to focus only on problem issues rather than looking for solutions to problems.

In our project, we chose to use the critique workshop as follow-up to the storytelling session. This gave the participants a rich field of ideas to begin their critique of current problems, and helped focus their brainstorming sessions. The critique addressed a broad range of issues such as:

- information loss;
- budget and pay orders are not tracked between departments;
- full capability of software not known;
- too much noise.

As in many American workplaces located in city office buildings, the participants were also frustrated by the lack of space, poor ventilation, and lack of privacy.

Fantasy Workshop

The structure of the fantasy workshop is similar to the structure of the critique workshop (i.e., brainstorming followed by group discussions and presentation) and it uses the same guidelines. In addition, facilitators try to encourage participants to:

- act as if no idea is too far out—anything is possible;
- be unrestrained by existing limitations or constraints, be they economical, technological or organizational;
- remember that the vision should be something that people really desire.

In the nonprofit organization's fantasy workshop, as in other fantasy workshops we have facilitated, the range of ideas was wide and exhilarating. Calls for a spa and relaxation room were heard, but the staff members also talked about:

- the need for access to all information except classified material;
- the need for an integrated communication system;
- the idea that perhaps they could find one tool that does everything;
- "a boot camp" where staff members could learn or relearn specific software packages.

Implementation Workshop

In general, the purpose of an implementation workshop is to initiate actions that bridge the gap between vision and reality. In setting up an implementation workshop some crucial elements for the participants and the facilitators are:

- being aware of obstacles, hindrances, and limitations;
- being aware of possibilities;
- bringing down to earth the visions stated in the fantasy phase;
- prioritizing issues;
- setting up a plan of action.

In the organizational setting of our project we expanded the implementation phase to include a Storyboard technique as a way of becoming more specific about proposals for a new desktop environment. The technique is inspired from the use of storyboards in film production. The use of storyboards in the design of computer systems can be a preliminary way to sketch plans for future applications (Andriole, 1991) and (Curtis & Vertelny, 1990). The technique is also related to techniques developed in the UTOPIA project in Scandinavia where they were used to help typographers design new tools (Ehn & Kyng, 1991). Although only a little time in our project was dedicated to storyboarding, the participant's storyboards, which included posters with drawings and "post-its," helped them express their ideas concretely. In the beginning of the session many of the participants were a little nervous about trying to express their ideas graphically, but once the work groups got together to depict their plans, the group process put the participants at ease.

The last part of the implementation workshop included setting up a plan of action. For each item the plan included what to do, when, and who should do it. Some of the actions were immediately initiated, while others were stated as suggestions that would go into the final report. One suggestion for immediate action was the simple, but effective idea that staff members could take a guided tour around the office and see the software that was available in other departments—for example, one department used Macintosh systems, while another had a limited Local Area Network. One staff member in each office was responsible for getting permission from his or her department head and for setting up a demonstration so that participants could get an idea of the other kinds of software available within the organization itself.

Vendor Visits

Vendor visits were planned in order to provide people with an opportunity to try out and compare specific desktop packages. We had looked forward to the chance to let the participants try some of their ideas out using packages that were available on the market. Yet this was the most unsuccessful part of the project, for the vendors that we contacted were not used to setting up the type of participatory sessions we had envisioned. Although large vendors are used to setting up sales demonstrations for managers, it appears that they were unfamiliar or uncomfortable with the idea of providing nonmanagerial users with an opportunity to get hands-on experience.

We did manage to offer a subgroup of participants a visit to a single hardware vendor who had invited different software vendors to take part. Although we explicitly asked the vendor to arrange a hands-on session, it turned out that they arranged a conventional demonstration with no opportunity for the participants to try the packages. We made a number of attempts to arrange other visits but were unsuccessful. In fact, we think that for participatory design to succeed in the United States, one of the major issues that needs to be addressed is the readiness of vendors to provide hands-on demonstrations to potential users. From our experience it appears that vendors tend to view managers as their potential market, steering away from "wasting time" showing systems to nonmanagerial users. Yet the case can easily be made that involving users in the selection process could lead to the opportunity for more sales as staff members try out new packages.

Reporting

The final project report to management was written by us in close cooperation with the participants. Our draft was carefully reviewed by the participants, leading to a detailed revision and improvement of the report itself.

As discussed in the implementation workshop, the participants planned how they would present their ideas to management. At a meeting between management, participants, and ourselves, the participants presented the specific ideas they found most important. Management's reaction was extremely positive, finding that both the results of the project and the participatory process were useful to the organization.

CONCLUDING DISCUSSION

From both the perspectives of management and the participants the goals of the project were achieved. According to the participants, the workshops were very helpful in enabling them to find out that other staff members had problems similar to their own and to develop strategies for coping with current problems. Additionally they became familiar with the fact that there were networks available that would suit their needs. From management's perspective, the project confirmed their original study about user needs and gave them *directional support* for making decisions about desktop packages.

But, the project also raised a number of potential problems:

We invited the MIS director to the opening workshop, but did not include him to any of the subsequent workshop activities because we wanted to avoid any "finger pointing" about the responsibility for current problems. As in many organizations, the MIS department was responsible for routine computer operations, and sometimes, as could be expected, there were antagonisms between other departments and MIS that grew out of preexisting problems. While a Scandinavian approach advocates direct involvement of system developers and participants, it is important to recognize the difference between the interests of system developers as outside consultants and the internal interests of an organization. In many cases outside facilitators can be helpful to bridge the gap between current problems and future planning.

The goals of the project included both giving *management* directional support and enabling the *staff* to better formulate *their* expectations concerning computer support for their work activities. This situation contained a potential conflict between the needs of management and the needs of the staff. The Scandinavian projects have mainly been in a context where facilitators have supported the staff. Workers in the Scandinavian trade union projects have been able to gain influence in similar situations basically either "by using good arguments" or "by using power" (for instance threatening to deny the use of new equipment)—the latter situation is supported by laws that allow union members to have a say about the type of technology used in the workplace (see Bjerknes, Ehn, & Kyng, 1987).

Obviously the American experience is different in both legal structure and workplace practice. Clearly participatory design strategies in the United States have to take into account the fact that the system design projects are initiated by management and need to meet the objectives of managers—both upper management and MIS management. It is probably more than a small change to begin to inject the idea that meeting worker's needs could and should be included in the decision-making process. In practice this is beginning to happen. At the Participatory Design Conference (PDC '90) and at many small informal gatherings, systems developers are talking about their experiences with supporting users to find ways to participate in decisions about computers in the workplace. The "good arguments" for supporting this approach have begun to gain acceptance. As in our project, many developers are showing that involving participants early in the design process and continuing staff participation help both managers and staff members to identify possible problem areas and to resolve misunderstandings before they develop into full-fledged design errors. Whether or not this approach can be cost-justified in traditional system terms remains to be seen, but intuitively we know that involving users as ongoing participants in the design process doesn't have direct costs and may bring about concrete results in both making decisions about computer support and in facilitating communication within an organization.

Based on our experience we suggest several issues that, we believe, need further attention both in the United States and in other countries where workers rights may not be explicated protected. In particular we feel that system development practices need to incorporate strategies for integrating user needs into management initiated projects. Our approach advocates using techniques, like the workshops discussed here, to enable staff members to voice their own concerns and gain experience and support in participating in the decision-making process. Clearly this can't be done in most organizations without some form of management approval and MIS involvement. Moreover the decision-making process needs the active and supportive involvement of vendors who are willing to let staff members get practical experience with new hardware and software. These last two areas represent changes in the way management views the system development process. These changes may, indeed, turn out to be anything but small.

Authors note: We thank the management and staff members of the organization we worked with for their enthusiasm and encouragement during the consulting project. Their motivation and eagerness to try new approaches were particularly appreciated. The title of this article was "borrowed from" a 1973 novel by Marge Piercy, which illustrates that small changes over time have a profound impact on the way we live and work. In fact, a central character in the novel was a computer systems developer.

REFERENCES

Andriole, S. (1991). *Storyboard prototyping.* Wellesley, MA: QED Information Sciences.
Bjerknes, G., Ehn, P., & Kyng, M. (1987). *Computers and democracy: A Scandinavian challenge.* Aldershot, UK: Avebury.

Bødker, S., & Greenbaum, J. (1988). A feeling for systems development work. In K. Tijdens, et al., *Work, work and computerization*. Amsterdam: North Holland.

Bødker, S., Greenbaum, J., & Kyng, M. (1991). Setting the stage for design as action. In J. Greenbaum & M. Kyng (Eds.), *Design at work—Cooperative design of computer systems* (pp. 139–154). Hillsdale, NJ: Lawrence Erlbaum Associates.

Curtis, G., & Vertelney, L. (1990). *Storyboards and sketch prototypes for rapid interface visualization*. Tutorial, Conference on Human Factors in Computing Systems. Seattle: ACM.

Ehn, P., & Kyng, M. (1991). Cardboard computers: Mocking-it-up or hands on the future. In J. Greenbaum & M. Kyng (Eds.), *Design at work—Cooperative design of computer systems* (pp. 169–196). Hillsdale, NJ: Lawrence Erlbaum Associates.

Greenbaum, J., & Kyng, M. (1991). *Design at work—Cooperative design of computer systems*. Hillsdale, NJ: Lawrence Erlbaum Associates.

Jungk, R., & Müllert, N. (1987). *Future workshops: How to create desirable futures*. London: Institute for Social Invention.

Kensing, F., & Madsen, K. H. (1991). Generating visions: Future workshops and metaphorical design. In J. Greenbaum & M. Kyng (Eds.), *Design at work—Cooperative design of computer systems* (pp. 169–196). Hillsdale, NJ: Lawrence Erlbaum Associates.

Computer Professionals for Social Responsibility

Computer Professionals for Social Responsibility is an alliance of computer professionals concerned about the impact of computer technology on society. Decisions regarding the use of this technology have far-reaching consequences and force us to examine our basic values and priorities.

As technical experts, we are responsible for providing the public with realistic assessments of the power, promise, and limitations of computer technology. As concerned citizens, we must direct public attention to critical choices regarding applications of computer technology where there are potentially dangerous consequences.

Members of CPSR believe that computer technology should make life more enjoyable, productive, and secure. We are working for a world in which science and technology are used not to produce weapons of war, but to foster a safe and just society. These concerns impel us to many forms of action:

- We encourage public discussion of and public responsibility for decisions involving the use of computers in systems critical to society.
- We work to dispel popular myths about the infallibility of technological systems.
- We challenge the assumption that technology alone can solve political and social problems.
- We encourage critical examination of social and technical issues within the computer profession, nationally and internationally.
- We encourage the use of computer technology to improve the quality of life.

CPSR traces its history to the fall of 1981, when several researchers in Palo Alto, California, organized a lunch meeting to discuss their shared concerns about the connections between computing and the nuclear arms race. That meeting led to the eventual establishment of CPSR as a national organization, which now has 21 chapters throughout the United States. CPSR also has three staffed offices: the main office in Palo Alto; an office in Washington, D.C., which serves as the center for our civil liberties and privacy program; and an office in Cambridge, Massachusetts, which is the headquarters for our work on priorities for research funding in science and technology. We are affiliated with similar groups in Canada, Australia, New Zealand, Great Britain, Germany, Finland, and Italy.

Throughout its history, CPSR has sought to use the technical expertise of its members to influence critical questions of public policy. Our past accomplishments include:

- CPSR published the first papers and held the first public debates on the computing aspects of the Strategic Defense Initiative, or "Star Wars." CPSR members also testified before a U.S. Senate subcommittee on the feasibility of SDI.
- CPSR/Boston produced an award-winning slide show and videotape called "Reliability and Risk: Computers and Nuclear War."
- CPSR members produced the first book for general audiences on the ways in which computers revolutionize modern weapons systems, Computers in Battle: Will they Work?
- At the request of a House subcommittee, CPSR studied the FBI's proposed National Crime Information Center upgrade (NCIC 2000). CPSR's report was widely credited for the FBI's subsequent decision to drop a proposal to track individuals who had not been charged with any crime.
- CPSR co-produced a "Special Report on Computers and Elections" for the 1988 Presidential Campaign, highlighting the potential for errors in electronic vote-counting systems.
- CPSR filed lawsuits under the Freedom of Information Act to force the FBI and Secret Service to reveal whether they monitor computer bulletin boards and electronic mail.
- CPSR has organized several national conferences, on such topics as the Directions and Implications of Advanced Computing; Computers, Freedom, and Privacy; Participatory Design; and Computers and the Environment.
- CPSR helped lead a successful grassroots campaign to convince the Lotus Development Corporation not to release their proposed Marketplace: Households product, which would have included data on 120 million Americans.

The national CPSR program focuses on the following project areas:

- *Reliability and Risk.* This area reflects our concern that overreliance on computing technology can lead to unacceptable risks to society. It includes, but is not limited to, our work in analyzing military systems such as the SDI.
- *Civil Liberties and Privacy.* The growing use of computers for record-keeping has brought with it the danger that the vast amount of information maintained on individual citizens threatens the fundamental rights of privacy. Centered in our

Washington office, the Civil Liberties and Privacy Project has been concerned with such topics as the FBI National Crime Information Center, the growing use of databases by both government and private industry, the right of access to public information, extension of First Amendment rights to electronic communication, and establishing legal protections for privacy of computerized information.

• *Computers in the Workplace.* By the mid-1990s, more than half of United States workers will use a computer on the job. The increasing use of computers in the workplace raises important social issues, which range from adverse effects on health from poorly designed workstations to the restructuring of the labor economy. The CPSR Workplace Project has concentrated its attention on the design of software for the workplace, and particularly on the philosophy of "participatory design," in which software designers work together with users to ensure that systems meet the actual needs of that workplace.

• *The 21st Century Project.* For many years, most U.S. research in science and technology has been funded by the military and directed toward military needs. With the end of the Cold War and the changes that have swept Eastern Europe and the Soviet Union, it is time to refocus our scientific and technological research toward the problems that society will face as it enters the next century. The 21st Century Project, which is led by CPSR from its new Cambridge office, is a coalition effort with other professional organizations to work toward this redirecting of national research priorities.

In addition to these areas, CPSR also has several chapter-based projects and national interest groups spanning a wide range of issues, including computers in education; community networks and access to information; computers and the environment; viruses and threats to computer security; computerized vote-counting systems; the status of women in computer science; opportunities for non-defense employment; and the implications of speculative technologies such as nanotechnology and virtual reality.

People join CPSR both because they want to have an effect on the world outside the organization, and because they are in search of a community of concerned people within the profession. Through our national offices and our local chapters, CPSR attempts to serve both interests. CPSR membership is open to all, and we encourage you to join with us to ensure that computers can be a positive force in our society.

For additional information, please contact

CPSR
P.O. Box 717
Palo Alto, CA 94302
(415) 322-3778

Author Index

Subject Index